Beyond the Glory

A Memoir of a Climax Miner

Jim Ludwig

The Climax Glory Hole and Mine-Mill complex 1973

Published and distributed by
The Pleasant Avenue Nursery, Inc
506 S. Pleasant Avenue
PO Box 1669
Buena Vista, CO 81211
Phone 719-395-6955
Fax 719-395-5718
e-mail   pan@amigo.net or jamesjlud@amigo.net

$17.95

Beyond the Glory Hole
A Memoir of a Climax Miner

Copyright ©2005
James J. Ludwig
First Printing 2005

All rights reserved.

ISBN: 0-9679419-1-1

Other books by Jim Ludwig
The Climax Mine ©2000
    An Old Man Remembers the Way it Was.

Printed in the United States by Morris Publishing
3212 East Highway 30
Kearney, NE 68847
1-800-650-7888

Dedication

This book is dedicated to Gary, Eleanor and the rest of my family who have nurtured me back to health and made this book possible, and to my siblings and friends who allowed me to include them in my memoirs.

I would like to make a special dedication to Dr. Gibb and his staff who quarterback the team that monitors the health of this eccentric old man. To Dr. Corenman and Dr. Peters whose surgery rescued me, to their staffs and the staff of the Vail Hospital. To Drs. Gaul and Fralick who keep a somewhat reluctant heart beating. To Lura, Tammy, the helpful staff of the Mountain Medical Center, and others too numerous to mention who make my life possible.

Also to Mary Lee Bensman for editing, Steve Voynick and Lynda Larocca, Bob and Joyce Renaux, Dr. Bill Simon, and Deb Ellis for advice, moral support and assistance.

Thank You! Thank You! Thank You All!

CONTENTS, Beyond the Glory Hole

Title and Copyright

Dedication

Title	Page
Chapter 1, The Glory Hole 1971	1
Chapter 2, The First Ten Years	5
Chapter 3, Growing Up	25
Chapter 4, Off to College	43
Chapter 5, At Climax Again, 1952	55
Chapter 6, Evergreen and Buena Vista	77
Chapter 7, Western Operations	97
Chapter 8, In Retrospect	105
Chapter 9, Reflections	109
Chapter 10, An Old Man Learns to Weave	129
Chapter 11, Deeper Reflections	141
Chapter 12, Media and the News	151
Chapter 13, Epilogue, 2004	157
Appendix, Glossary	i
Sources	v
About the Author	vii
Plaudits	ix
Shipping information and Order form	xi

b

Chapter 1, The Glory Hole 1971

"That, my friends, is the Glory Hole!"

The men all stopped and stared at the sharp escarpments outlining the donut-shaped depression so obviously unnatural on the gentle slope of the mountain. It extended nearly a mile across and a thousand feet in elevation.

The shift boss continued, "Every dollar, every ton of rock, every pound of Moly that built this place has come from that hole. Many people depend on that hole for their livelihood and life-style. It has taxed the ingenuity of men, and inspired mining methods and equipment previously unknown. The Glory Hole symbolizes Climax, identifies Climax, and in fact is Climax. If you stay long enough, someday you too will proudly become a part of it."

He was leading a half dozen new hands from the Fremont Trading company store on their way to a day of indoctrination and safety training for a job in the Climax mine. Each man had been extended credit to purchase work clothes, a lunch bucket and lamp belt to begin his work in the mine. He would be trained for a week and hammered with safety rules before being assigned to a permanent crew.

The store was located across the highway from the gate that was the entry to the mining camp and the town of Climax. As the men looked to the east, above the jumble of houses, mill buildings, conveyors and railroad tracks, the Glory Hole was such a dominant feature on the face of Bartlett Mountain that invariably someone would ask what it was. Anticipating the question, the boss had honed his reply on countless new employees he had trained. He urged the men into the Suburban that had been outfitted as a makeshift bus and they were off to their first day of work at the Climax mine.

The new hand ritual had happened twenty years before in 1951 and I now stood by myself in a parking lot ringed by explanatory signs for the tourists. It was the exact place that the store had been.

Little powder puff clouds cast fast moving shadows up the slope of Little Bartlett, down into the glory hole, up across the central core, down into the south block cave, up the escarpment onto Ceresco Ridge and scurried off the peak of Mount MacNamee into the fairyland where the shadows of all clouds go. I leaned against the fender of the Blazer I was privileged to use, studying the moving patterns, looking for any slight change to the surface I had studied a few days ago. Habitually, I took the Copenhagen can from my shirt pocket, tapped the lid, opened it and took a pinch of snuff. It is a nasty habit; I'm going to quit it some day.

Just like the shadow of a cloud, I had blown in from the north, lingered a while in the hole and now was about to struggle out and beyond, only to vanish into the world, where millions of such shadows and lives appear and disappear each day.

Twenty years before I had wondered about the Glory Hole. I went to work and learned damn well how it came to be, and I became a part of it, and it a part of me. It was a huge, cavernous eye, looking toward the heavens from the face of the earth. It had become my eye to the world. I had watched it grow and grow, and I had grown with it.

Years later, I had studied aerial photos of that same glory hole, trying to determine the surface indications of a tremendous point of pressure crushing our workings underground. I had watched the cloud of dust rise from the celebrated blast of 1964 when we peeled some high grade from the central core.

This day in 1971 I had been asked to move beyond the Glory Hole, to manage other places, engineers, prospects, and mines.

I didn't want to go. I wished they hadn't asked, but I would go anyway.

I didn't realize it then, but the very soul that made Climax and its Glory Hole so special had already gone. It had moved down the hill, a bit at a time, with each family that moved away. Each house that moved to West Park made Climax a little less special until finally when the Store and the last boarding house closed it became just another place to work, another job worth so much per hour and no one really cared if the Glory Hole had once been an inspiration to young men like myself.

So I would leave, the last of a dying breed that had cut my managerial teeth while living in Climax. We had watched the Glory Hole each day as we walked to work and looked for change each time we came in the gate on our way home. No longer would anyone awaken in the night if the trains ceased rattling down the track, and to whom the sparking of the trolley poles was as reassuring as the northern lights are to Eskimos.

Later I would watch from afar as an open pit smoothed and benched the edges, destroying the grandeur of the man-made canyon. Later still, I would watch the place grow silent, the flock of ravens riding on the updraft at the Glory Hole edge was then reduced to one or two because the garbage that had been their food source had disappeared.

I've been retired for more than twenty years, but the goose bumps rise as I drive by, and ancient "What if?" and "I wonder why?" questions again cross my mind. I was a farm boy from Wisconsin who somehow made it here, who had become a part of the Glory Hole and then moved on. You cannot really leave, for its life, its people, and its lessons will be with you always.

I would like to share with you some tales of where it all began on the farm, more tales of the mine and beyond, and some of my personal philosophy, before this shadow of a cloud called life disappears completely into the netherworld.

The game of life is nearly over, but what a game it has been.

Chapter 2, the First Ten Years

The Farm

My Mother once told me, when my wife Mickey was pregnant, "You may not wish for another child, but when it comes, you will love it just as much as all the others. That is what God intends." So when my Mother was expecting another baby, her sixth, in the fall of 1928, I'm sure she prayed that she could follow through on God's intentions. She didn't realize how hard it was going to be with this particular one. If she had, she might have done what my brother-in-law Buck advised about any unwanted birth on a farm, "Hit it on the head and feed it to the crows."

Pa probably could not, or if he could, would not explain his deepest feelings about having another child to anyone except his wife. I can almost hear him joking to his friends that he was glad the baby had arrived before deer hunting season, so Bertha wouldn't have to worry while he was up north in the woods somewhere, and she was home alone.

Betty, not yet four, and Art at six were probably most excited to have a new baby, while Ruth and Lucille, nine and eleven, just absolutely loved the new baby as only little girls can. Lucille tells me Mother included them all in such things as giving the baby a bath, and thinks this is one of the reasons our family has been so close over all the years. Harold, at fourteen and trying to be grownup about it all, pretended he could care less.

The kids were allowed to pick the baby's name from a list made by Mom. They chose "James," and Mom agreed, if they would promise to never call me "Jim" or "Jimmy." Lucille said that was all forgotten by the time I went to school.

Lucille thinks our Grandma Ludwig, who was a midwife, delivered me right in the farmhouse, although my birth certificate is signed by E. B. Elvis, M. D. in beautiful, flowing script. Vincent Jakel, registrar, recorded this birth as occurring at 7 PM, Oct. 29, 1928, in the State of Wisconsin, County of Taylor, Township of Little Black.

I don't know when I began to remember events, sometimes just simple incidents, sometimes just emotions, or sometimes a mother's love and attitude, but they became so deeply ingrained that it is hard to start a story without first referring to them. One of my earliest memories troubles me a bit.

It was the fall of 1934. Donnie Gebert and I were on our way home from school. Just first graders, we were excused a half hour early, to walk home on a pleasant autumn afternoon. Throwing rocks at fence posts, we were in no hurry.

We were leaving Liberty, a two-room red brick school. A mile and three quarters to the west was home. We walked it every day.

As we passed Ryner's cheese factory, a man was sitting near the road, leaning up against a tree in front of Ed Oelke's house. He arose and silently fell into step with us. I had no idea who he was and it was obvious that Donnie didn't either.

"Hi kids," he said. We shyly answered, "Hi."

Then he was silent for a while. We walked along.

"Do you know Otto Will?" he asked.

"He lives right there," said Donnie, pointing to the next farm on the south side of the road.

His voice trembled just a bit. "I was in the war with him." He paused and stared intently. "You're sure that is where he lives?" A wild look came to his eye, "I'm going to kill the son-of a-bitch!"

Now I was scared, as scared as I have ever been. I broke into a run, my lunch pail swinging wildly about.

"Hurry Donnie, hurry!"

"Wait for me, wait for me!"

Donnie soon was home. I had a quarter mile to go.

Ma was peeling potatoes, sitting at the kitchen table when I burst in.

"Mamma, Mamma, a man is going to kill Otto Will!"

"I don't believe it, Jim. You're not making this all up because I see you wet your pants?"

I didn't know that I had wet my pants. "Cross my heart and hope to die," I said. I told her the whole story.

"Well, we'll tell Pa when He comes in. Now change your pants and wash your hands and take off your good school shoes," she said.

She wasn't too concerned and neither was Pa when I told him. He looked knowingly at Ma, "That must have been old you-know-who. I heard he was in town. He was gassed in France during the war while Otto was his sergeant there. Otto can handle him."

"Jimmy's a cry baby, Jimmy's a cry baby." Little sister Marianne joined in the fray.

A reality of life sank in. I never heard about the incident again. A father's reassurance made a threat of murder no big deal. It seemed to say "Don't worry son, someone will handle it."

Life is like that. It always seemed to be a lot of fun for me, but it was not for our folks. They struggled into the depression with too small a farm, too many kids, and then Pa had a severe handicap when he lost the fingers on his left hand in a silo filling accident. Cows are hard to milk with one hand.

I find myself often returning to those early days when telling a story, for they set the course for my life. It is interesting to note that the things you remember clearly were those things that seem to define the limits of your emotions. In particular, the attitude toward animals, nature in general and pets became clearly defined and has not changed in seventy years.

I was in the first grade, and came home from school one night, expecting to be met at driveway by my dog, Lad. When I arrived at the house, I asked Ma if she had seen him. She said, "Yes, but they had to shoot him and Pa buried he and Patsy out behind the barn by the manure pile." It seems that two dogs had killed some of Julius Peissig's turkeys the night before, and old Patsy had also elected to return home from Uncle George's. Someone suspected them. Killing a poor farmer's livestock was a mandatory death sentence.

Several years later, the dogs in our neighborhood began to band together at night and chase cows. Very early one morning Pa called both Art and I out of bed. That was unusual because ordinarily his call was, "Jimmy, time to get the cows!"

He explained, "Last night about two-thirty I heard Anton's cows and some dogs raising a ruckus, and then Anton shot a couple times. So I got up, grabbed the deer rifle and went outside. Pretty soon here came old Pete running across the field from Anton's, so I had to shoot him. Will you kids drag him back to the cow pasture and bury him?"

Most of the dogs in our neighborhood were dead before the trouble stopped. I vowed I would never have a dog again.

Times were really hard, but 1936 was a bad year by any measurement. Pa had lost his fingers while chopping corn the previous fall. The income from the cows barely kept food on the table, and would not pay the doctor bills or the mortgage payments on the farm.

Pa decided to plant a cash crop of peas to be sold to the cannery in Medford. Planting and cutting times were determined by the cannery field man. When he told Pa they wanted the peas at the viner the next morning, Pa hitched the horses, Dick and King, to the mower and started cutting after dark so the peas wouldn't wilt before loading. It didn't matter that it was too dark to see. The horses knew where to walk after the first round. When the peas were all cut they were hand-loaded onto a wagon to be hauled to the viner. At the viner the fresh peas were popped from the shells for transport to the cannery.

Pa, Art and some neighbors worked all night and I had just driven the cows from the night pasture for the morning milking when the first load came from the field for inspection. It was covered with big green, orange spotted dill worms. Millions of them! The field man could not accept the peas. The crop, the seed money and work were all lost. There was to be no cash income, and Pa never made another payment toward the mortgage on the farm.

Later that fall, Harry Ryner needed a man to run a logging camp west of Ogema, and he asked Pa to round up a crew of local lumberjacks with necessary horses and gear. Pa was happy to oblige, and was ready to go at freeze up that winter. So you can imagine how excited I was when the folks decided Lucile, Art and Betty could handle the chores on the farm and that Marianne and Jimmy could spend Christmas vacation at camp. My mother was the camp cook.

Pa's instructions were simple. "Don't get lost in the woods and stay out of the way so you won't get hurt." That gave me a lot of latitude.

There are some impressions I'll never forget. One was the warmth and smell of food and bakery in the cook house, especially compared to the working man smell of the bunkhouse, with the wet wool clothes hanging overhead, soaked with sweat and snow. Another was the camaraderie of the lumberjacks as they played cards in the evening by the light of a kerosene lamp. The game often paused at a particularly loud snore from some exhausted lumberjack down the row of bunks or to tease Louie's little boy.

The woods are incredibly silent when it is below zero and there is no wind. There would be a muffled thump when an evergreen branch would decide to drop its snow load. Only the squeak of four-buckle overshoes on the cold snow would break the silence. Even the Canada Jays flew silently about their camp robber business. I would strain my ears trying to pick up the sounds of the workers, but except for the occasional far off sound of a tree crashing to the ground, even the men and horses worked silently. The two-man cross cut saws cut with a whisper and sleighs glided silently over the snow, sometimes a tinkling of the trace chains the only sound. From a distance the sound of a razor sharp double-bitted ax notching a tree for felling was simply a rhythmic "thuk---thuk---thuk".

I loved it, and I still love to work in absolute silence.

I remember well the "fog horn" Pa would use to call the men from the woods. It was about seven feet long, made of tin, tapered it's full length from a simple ring mouthpiece to about ten inches at the bell. By pursing his lips and blowing he could make a low-pitched sound that would reverberate and echo for miles through the woods. Soon the lumberjacks and teamsters would come filing in.

Table scraps were placed outside the kitchen windows for the birds, and under the tutelage of my mother, I became an avid bird-watcher. She taught me to observe the unique actions of each species so that it could be identified in flight or when perched, even if the exact color was not visible.

That winter was a memorable time for me. It was just a lot of hard work for my parents. Financially, the job did little to stave off the inevitable loss of the farm.

Being poor is a state of mind, not of material. My folks would simply not consider the fact that we had so little to be a disadvantage. Rather it was an opportunity to wish for and work for a better time. If they did complain or blame someone else for their misfortune, we kids never heard it.

I learned to love classical jazz at an early age. The following excerpt is from my eulogy to my oldest brother Mickey:

"The Harmony Kings?" How proud I was to have a brother who really had a dance band. The farmhouse kitchen was used for practice and it was rather small. I know the drummer had to set his instruments so he could sit in the doorway of the bedroom, with the base, snare drums and cymbal placed next to the kitchen stove. I think Mother's pride and joy, the old upright piano, was actually in the living room. Mickey, Mike and Tom Habermayer were going to practice and I had tucked into a corner to listen when Tom asked me if I wanted to play the drums until Lawrence Etten arrived. Oh, did I ever! Tom perched me on the chair, gave me the brushes and showed me how to beat once with the left and twice with the right and not too loud. My feet swung high above the base pedal and once I bravely hit the symbol with a brush at the end of a song.

I was thrilled and the words from the song "Cheek to Cheek" echoed through my mind; "Heaven, I'm in Heaven, and my heart beats so that I can hardly sing."

Harold became "Mickey" after he and some of his teenage friends saw the Walt Disney cartoon "Steamboat Willie" that introduced Mickey Mouse playing the piano. The nickname never left him. Every time I hear Duke Ellington play the piano, and that is often, I hear Mickey sliding that cord note down the bar just enough to make a distinctive jazz sound. Today, I can listen to Duke, Ella Fitzgerald or Louie Armstrong by the hour. Mickey played the accordion all his life and at age 80, he was able to play for my second wedding.

Pa, Mickey, Art, and Jim. 1933

Let me recite a short fictionalized story of an event that might have happened. I know I don't remember living at home with Mickey.

Why Mickey Left the Farm

I woke up and heard the singing even before I heard the car. "Sw-wee-et ---Ad--oo--line" That had to be Tony Stulgo's baritone, which was then echoed in perfect three part harmony. "Sw-wee-et--- Ad-- oo--line." That had to be Mickey, Tom and Mike and a trill on the harmonica, they must have had Buck with them, too. "My Ad--o--line," - My Ad---oo--line," trailed off into Tom's base. "At night-- dear--- heart, at night --dear ---heart, -------alone ---I pine, alone ---I pine."

The car slid to a stop in front of the house. "In all my ---dreams –" A door banged, "In all my--- dreams,"

I recognized Mickey's voice saying, "Jeez Tony! I don't mind if you take a leak on your own car wheel like a dog, but do you have to do it in the snow in front of Pa's house?"

The singing continued, "Your-- fair-- face--- gleams, your-- fair-- face---gleams."

Tony said, "Quit your belly aching. You didn't want me to wet my pants, did you? Then Tony asked, "Hey is there anything left in that bottle?" "You're --the-flower--- of ---my--- heart, --of ---my--- heart, SSuh --wwe---eeet ----Ad---ooooo ---line."

Tony answered himself, "Deader than a doornail, huh? It was good while it lasted."

Mickey interrupted, "I'm getting sick, let me out." Tony went on, "Can't handle your booze, eh kid? Hell, your damn near eighteen years old."

I heard footsteps running toward the outhouse, a car door slammed and wheels of the Model A Ford spun gravel across the yard. "Suh--w-eet--- Ad--o-line, Sw--eet Ad--ol---ine," echoed off the barn as they went down the driveway. I pulled the quilt up around my ears to go back to sleep.

"Art, Jimmy, wake up!" Pa shouted up the staircase. "Mickey set fire to the outhouse, get up and help me put it out." The door to the stairwell slammed, "Ma, wake up and get the milk pails, I'm going to start the engine on the pump."

I jumped out of bed -- Art was already pulling on his overalls. I looked out the south upstairs bedroom window, buttoning the back flap of my wool underwear as I went. There was a blaze all right, the flames flickered little shafts of light on the bedroom ceiling as I pulled on my wool socks, heavy shirt and overalls. I put on my four-buckle overshoes at the back door.

Ma was hustling to get the kerosene lantern lit and build a fire in the kitchen stove.

That outhouse was Pa's pride and joy. He had scrounged some old barn wood and built it about forty feet south of the house and west of the pump house so Ma wouldn't have to run clear across the yard in the dark of night to the two-holer east of the granary. He even painted it with some dark red barn paint and cut a crescent moon in the door. It only had one hole, but that was enough because it was for Ma, not us kids.

The last quarter moon cast an eerie glow on the drifts of dirty snow in the yard as I stepped outside, carrying a milk pail Ma had set out.

"Pop-a-diddle-diddle, Pop-a-diddle-diddle, POP-a-diddle-POP-a-diddle-POP-POP-POP-diddle-diddle-diddle." The little John Deere one lung engine settled down, "Pop-a-diddle-diddle, Pop-a-diddle-diddle," and Pa set the idler onto the flat belt, the pump jack screeched and began lifting water from the well and pumping it toward the galvanized tank upstairs in the house. "Scree-kaw-rattle, rattle, Scree-kaw-rattle, rattle," just like Harry Paul's old Guinea hen.

I ran toward the toilet, the fire was sort of around the corner of the pump house and I saw the outhouse itself was not on fire but a pile of kindling wood stacked next to it had flared up and shot flames nearly as high as the pump house

"Get out of there, you damn fool," Pa shouted.

"I can't, I'm so sick and my head is in the toilet hole," Mickey blubbered.

"Art, help me tip this thing over backward so it doesn't catch on fire," Pa said as he pushed and grunted.

"Damn, it's frozen down. Jimmy, bring me that pole and wedge it against that popple tree to break it loose."

Ma came with a pail of water, gave a mighty heave, missed the fire and hit Mickey who was kneeling inside the open door with his head hanging in the toilet hole, his pants down around his ankles.

Art came running around to help me with the pole, we wedged it in like Pa said, gave a mighty push and tipped the toilet forward instead of sideways, right on the door which slammed shut on Mickey's legs and pants.

I heard the neighbors, Anton and his son Franklin, come running across the field from their farm, carrying a couple of milk pails to help put out the fire. They yelled at Pa, asking where they could help. In the darkness, the first thing Franklin did was run into the now uncovered toilet pit. He was cussing his bad luck when he looked up to see Mickey's head sticking through the toilet hole, and let out a guffaw.

"Let me help you out of there," he said, and gave Mickey's head a push.

Mickey moaned, "You can't. My pants are around my ankles and the toilet door slammed shut on them when it tipped over. I can't move."

Pa said, "Let the dummy stay there and help us carry water to put out the fire in the woodpile before the pump-house goes up in flames." With five of us running with pails of water, although I could only handle half of one, and Ma turning the valve off and on, we soon had the fire under control.

The upstairs tank had pretty well emptied, so Pa just let the engine run to fill it back up. "POP-a-diddle-diddle- Scree-kaw -rattle, rattle, POP-a-diddle-diddle, Scree-kaw -rattle, rattle."

Anton and Pa rolled the toilet on its side and got Mickey's pants loose and pulled his head out of the hole. He was shivering and chattering from the cold and was too drunk to walk. Ma came running with an old blanket and wrapped it around Mickey and helped him to the house as Art was drowning the last embers from the fire. Franklin was scrubbing off his clothes and boots in a snow bank with an old barn broom. I was getting really cold, about half wet and getting over my fright, when Pa looked at Anton and said, "Let's go in the house and have a cup of coffee before it is time to milk the cows."

Mickey was bundled up before the open oven of the old wood stove when we came in. Ma had built a roaring fire and the old enamel coffee pot was boiling furiously. I took off my wet overalls, hung them behind the kitchen stove and stood in my damp underwear as close to the stove as I could. Pa was becoming angry all over again.

"How in the hell did you get into such a pickle and damn near burn our place down, young man?" Pa asked as he poured coffee for the men.

Mickey stuttered and shivered in his drunkenness, then through chattering teeth began.

"You know that Hartman from Dorchester that was married in Spring Green last week? Well, some of his Buddies decided to shivaree him for not having a wedding dance. We were in Jake's tavern having a few beers after practicing at Danen's last night, when this guy walks in and says, 'I'll give ten bucks and all the beer you can drink if you will come out to Lusher's dance hall and make some music for a shivaree.' Tony looked around at us, and then held out his hand. The guy slapped a ten-dollar bill in it and walked out the door. Tony stuck the ten in his shirt pocket."

"I mean that was a party. All of Dorchester was there and a good share of the Polacks from Lublin, her hometown. Somebody told me the groom had to pay for fifteen halves of Leinenkugel beer and the music. We were all feeling pretty good when Artie Woltz from out on Highway 64 came up to the stage. He asked Tony if he wanted to buy some moonshine. Tony reached in his shirt pocket and gave him the ten-dollar bill.

"In a while Artie came back with a nearly full gallon jug of clear booze and set it behind Tony's music stand. 'Where in hell did you get that?' Tony asked.

'You know my Pa, Louie? He rented the hay barn to some guys from Chicago for two hundred a month. Pa had never seen so much money. One night a truck pulled in and they unloaded a bunch of machinery that actually was a still. There were also a couple tough looking guys packing Tommy guns. When they got the mash cooking, you could smell it clear down by the Black River Bridge.

"This afternoon Pa went out and asked them what they were doing. They sort of pointed the Tommy guns at his head. Then another guy came over and handed Pa two gallons of this stuff and said, 'you keep your mouth shut or --,' and drew his finger slowly across his throat.

"Louie was so shook that when he came in, he handed me the jugs and told me to get rid of them. This is one, we tried it a little but it is horrible, you can have the rest. A guy stopped me in our driveway, pointed a gun at me and took the other one. So much for honor among thieves." Artie ambled off and was soon dancing up a storm.

"Mike said, 'we better try one too, Tony, before it spoils.'

" 'If this is what I think it is, it can't spoil any worse than it is, we better wait until we quit playing,' said Tony. When we finally played the last song, Tony opened the jug and passed it around. Talk about fire! I couldn't find my butt with both hands by the time we packed up our instruments to leave."

Pa interrupted his story, "What did all that have to do with starting the fire?"

"Well, we were singing in the car and I started to get sick when we pulled into the yard. I ran for the toilet, pulled down my pants, sat on the hole and tried to light a cigarette. Then it was obvious I had the wrong end in the hole and I dropped the match. It fell into the Sears Roebuck catalog we used for toilet paper and it caught on fire. In the struggle to get my head over the hole I kicked the burning catalog out the door and into the pile of kindling. I was nearly puking my guts out and didn't give a damn what was on fire. Then you guys came hollering and screaming and tipped the toilet over on the door, pushing my head through the toilet hole and pinning my pants tight around my ankles so I couldn't move. Then

Franklin tried to push my head back through the hole and that made it worse. I don't know who dumped the water on me. Oh, am I ever sick!"

Ma had helped him put his feet into a dishpan of warm water, 'so he didn't just die,' Ma said. Anton, Franklin and Art were laughing and Pa was becoming angrier by the minute.

"Jimmy, better go to bed," Ma said.

"I can't, my long johns are still wet." I would have said anything to hear the tongue-lashing Pa was about to give Mickey.

"Well get some dry ones."

"I don't have any."

"You do too."

"They have big holes in the knees."

"Put them on anyway. Off you go."

Ma was even more certain I shouldn't hear what Pa had to say.

So I ran upstairs, changed my underwear and pulled the quilt around me as I lay on the floor and looked down through the floor heat register. I peeked down just as Mickey in his bare feet bolted for the door to find a place to heave something from an already empty stomach.

Anton and Franklin conveniently decided it was time to milk their cows and Art left with them to start feeding ours. Pa followed them out to where Mickey was leaning his head on his arm against a tree. He took Mickey's other arm, pulled it across his own shoulder, around his neck and helped him back in the house.

I saw the two of them come in, Ma holding the door open, Pa almost carrying Mickey and I could feel the fight and the anger drain from my dad and near despair set in. My mother began to cry softly. Pa helped Mickey sit on the bench at the kitchen table. I think he passed out.

"What are we going to do, Bertha?" Pa asked quietly, as he took Mom in his arms. "We can't afford to keep him here if he just plays in that band. He has to work somewhere." "I don't know—I don't know." She said. "Help me put him to bed on the couch." She looked up at the register. "Jimmy, please go back to bed." You don't fool

Moms for long. She was standing, sobbing in Pa's arms when I crawled back into bed.

When I woke up, Mickey was gone. He had left in his old 29 Chevy with the few things he owned. My folks never spoke about it again. A few weeks later, Art told me that he had talked to Mickey at a dance in Chelsea. He was still playing with the band.

I became determined that I would go to work to support myself as soon as I could, and never be a burden.

jimmyj 3/01/03

As an old man the memories are clear, I recalled the events and emotions and was prompted to write;

On Being Lucky

A poor kid on a farm has the opportunity to learn that a lot a modern kid will never know.

You learn where babies came from, how they started in the first place, too. You learn that sometimes babies died, and that pets aren't really human, and sometimes they died too.

To take care of the animals was a duty from which there's no relief, it must done. There is no acceptable excuse.

You learn to shoot, then skin and clean a rabbit with your pocketknife. -- a knife you learned to sharpen, and still do.

The best of toys are made by hand; the loveliest of flowers are picked from the wild.

You learn that garden plants grow better when the weeds are gone, and so do friends.

You listen to the whisper of the wind, hear the sound not made by falling snow, you feel the rain fall gently on your flesh, you hear the music of the birds within the silence of the woods.

You learn how to be alone and not be lonely, to be in the dark and never be afraid.

How lucky to be born a poor kid on a farm.

jimmyj 2/20/98

My mother's influence was strong. She didn't hesitate to discipline although at times it consisted of, "Wait until Pa comes home!" The following bit is from an incomplete unpublished story, "I remember Buck Riemer," who was my brother-in-law.

I was sitting in the dirt, pushing a block of wood as if it were a truck, building roads between the weeds and singing to myself.

"I had a dog, his name was Jack.
He s--- all over the railroad track.
The train came by, the s--- flew high,
and hit the conductor square in the eye.
O-o-o-oh, it ain't gonna rain no more, no more...."

A rather large shadow from a rather large mom interrupted the lyrics, "Jimmy! Where did you learn that?" I stuttered a bit, "From Galen." "Well, where did he learn it?" "From Buck." I replied.

"I don't ever want to hear you sing that song again!" She said as she stomped off toward the garden. Her disapproval was quite obvious.

So at the tender age of six, I learned that you had to be quite careful where you repeated what you learned from Buck or Galen. I continued the tune anyway, quietly, under my breath,

"O-o-o-oh it ain't gonna rain no more, no more. It ain't gonna rain no mo-o-re.
How in the hell can the old folks tell? It ain't gonna rain no more."

jimmyj 2/16/98

The following story gives an insight into my Mother's family who all lived within close proximity during the early thirties.

How Galen Pet the Lamb

Aunt Lucy lived in Little Black at the east end of the Peissig road, about a half-mile from the Soo Line tracks. Uncle Eddy lived just down the road, and we lived west a couple of miles. Aunt Hattie lived in Stetsonville, a few miles to the north, so why not get together on a

Sunday afternoon to play some cards and drink a beer or two?

It must have been in Thirty Four, I know that Uncle George was well, so I was six, and Herbie too. Galen was eight, I think, and Art was twelve. There were a lot of cousins; most were girls and babies, which really didn't count to boys.

We guys fooled around the hay barn, played some ball, then Teddy said, "Let's walk up the tracks to Meyer's meadow, they have baby lambs that are so cute."

There were six of us, Merlin, Teddy, Art and Galen, Herb, and me, no girls allowed. What a thrill it was to walk the tracks, take big steps to hit every other tie, or walk a full rail length and not fall off.

Sure enough, there was a flock of sheep and we could see the lambs. We little boys crawled through the fence and started toward the sheep, I wondered why the big guys all stayed back upon the gate.

Galen, who was faster, ran on ahead of us and had reached out to pet a lamb, when the old buck ran up and butted him flat. Herb and I stopped dead still. I was scared. Galen got back up to run; the buck just knocked him down again. Herb and I had turned, when Merlin hollered, "Walk, don't run!"

The big guys on the gate were doubled up in laughter, urging Galen on, "Don't let him scare you, Galen. Just stand up and look him in the eye, then slowly back away."

I guess it would have worked, except that Galen couldn't stand there, lost his nerve and when he turned, the buck would flatten him again. Herb and I were scared and ran; just beat the buck back to the fence. That gave Galen just a start. He didn't make it all the way, the buck caught up and knocked him flat again.

The big guys told him, "Just lay there, pretty soon he'll walk away and go back to the flock."

Now I was safe and pretty brave. I laughed at Galen, too. We couldn't wait to get back to the house and tell the grownups and the girls how Galen was knocked down.

We've told this story for sixty years. We are all alive except Merlin, but this is the first time I have written how Galen pet the lamb.

jimmyj 4/3/97

I sent this story to Galen via e-mail; this was his reply:

"I read your poem. every word; The most bullshit I've ever heard.

You've milked that venture all your days. You think exaggeration pays?

Though I may be pushin' seventy-two, my memory rings clear and true.

Your words made me look like a dunce; cuz I was only knocked down once!

Do you recall by any chance, that that's the day you wet your pants?"

Galen
July 1997

I showed this story and Galen's reply to Art. He said the ram knocked Galen down at least three times.

To which Galen replied, "Art's memory sure has been slipping of late"

jimmyj 7/15/97

By the time I was eight years old, I had my own set of chores, and then the season specific work, such as shocking grain. When the oats ripened, it was cut and bundled by a grain binder. The scattered bundles had to be stacked in shocks until the thresh machine was available. Pa owned a thresh machine which he pulled from farm to farm with a steel wheeled McCormick-Deering tractor, which had a flat belt pulley to drive the thresher. The farm neighbors worked together bringing their teams and wagons to each farm in turn. We kids also gathered and had to help as soon as we could by driving the horses, running errands or sacking grain. Men carried sacks of grain into the granary for storage, while others pitched bundles into the thresher, loaded the wagons in the field or stacked the straw.

It was a real neighborhood affair. Wherever the rig finished in the evening, the farm wife was expected to furnish a meal, and often a keg of beer was tapped before dinner. There was a lot of opportunity for kids to have fun and get in trouble. In our dairy farming country, after the day was over, everyone had to go home and milk the cows.

Much the same procedure was used for silo filling with chopped green corn to make silage. Pa also owned a silo filler and it was that machine which chopped off his fingers.

Putting up hay was generally a single-family venture, and it meant kids always had work to do. However, our parents really believed the old adage that "all work and no play makes Jack a dull boy."

We would swim in Henry Paul's gravel pit or the Black River and fish in the local creeks. Pickup softball games were always on and the now unknown games of Blindman's Bluff, Kick the Can, Ante, Ante Over, or Long Dutch were played, depending on who was available. Today's organized sports were unheard of.

I had jobs helping farmers with their chores by the time I was ten, generally for little more than room and board.

My parents always encouraged me and my siblings to read, study, and learn whatever possible. I needed little encouragement, and always had a book in my hand. This trait prompted Buck to remark, "Jimmy will never amount to anything. He is just a dreamer."

Reading

When I learned to read another life began, the world grew and grew.

I identified with Peter Rabbit, dreamed with Robert Louis Stevenson. Met the monsters of my day while sailing with Ulysses; for a thousand and one nights I rode a flying carpet with my guide, Aladdin.

Rudyard Kipling told me how to be a man, alerted me to thieves and knaves. In the land of Hiawatha I saw shining big-sea water; watched the stag in Scotland, who at eve would drink his fill. I watched the great white whale, Moby Dick, and rode with Lancelot and Arthur; at the table round I ate my fill.

Zane Grey took me west and talked of men and horses, and the purple sage. Jack London told me of the dog, White Fang. Robert Service took me north with tales of the cold. Mark Twain told me all about the river, took me to the mining camps, and told of the lust for gold. I met men and boys, heroines and heroes, all across the globe, yet never left my home.

I would have been quite happy to be a student all my days, and in a way I've been. The only thing that interfered was that I had so many other lives that took my time. The early, necessary skills did not come quite as quickly without calculators and computers and the TV that you have today. Flash cards, paper, pencils and a blackboard are not quite as fun to do. There were only books, and books, and books.

From books I learned that other men had learned a lot; of medicine and architecture, mathematics and geography, engineering and religion, of science and the minds of men. Ancient history fascinated me, still does. Business, economics and the working world remain a daily study course for me. To know the thoughts of learned men is my desire, but I know just a few. Yet to know them and not help teach others is a selfish thing to do.

I sit at this computer, trying to make words say just what I mean to say. This is my current study course, English Composition 101. I'm glad that school never has let out, it never will, not as long as I can see this screen or hear a spoken word.

And there was music, always music. I learned that music could change your mood, bring happiness to your idle hours. My talent was so little, yet my love so great, that after seventy years my life still beats to the swing of classical jazz.

jimmyj 2/16/98

The Old House

1/16" = 1'

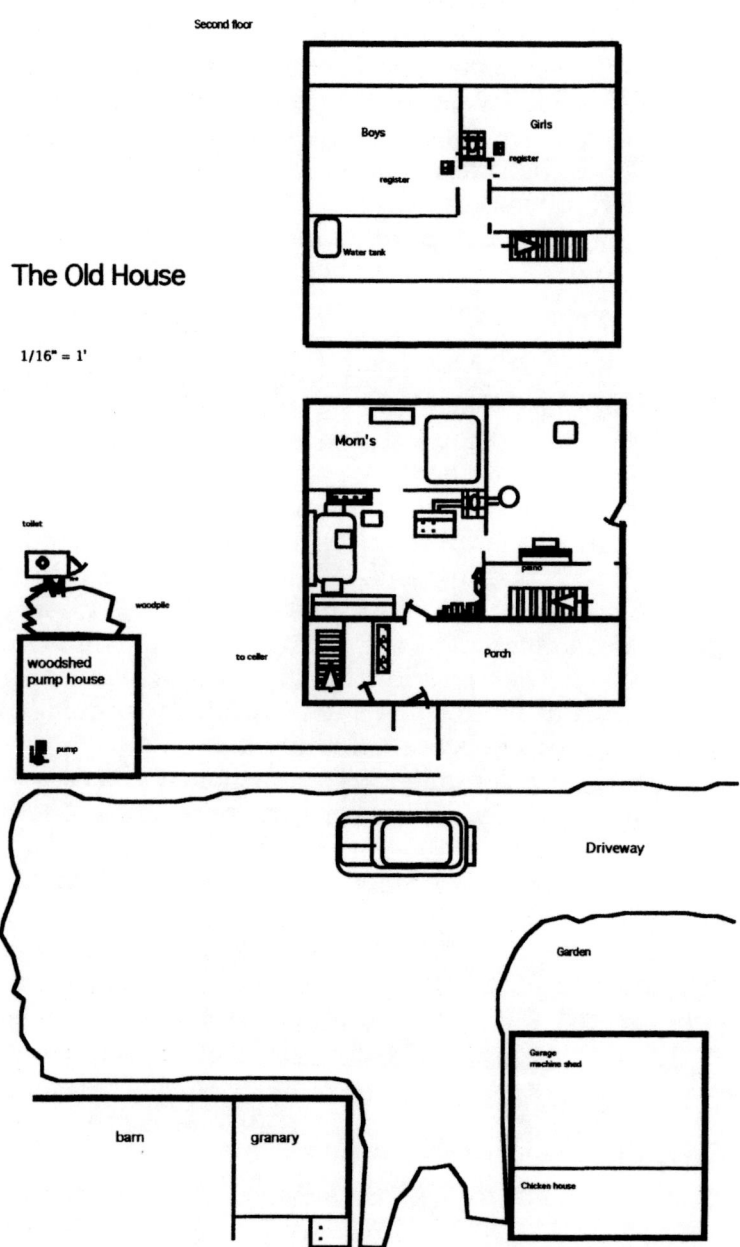

Chapter 3, Growing Up

In 1938 we moved from the place where I was born to a house in the northwest corner of the township of Little Black. I started the fifth grade at the Little White School that fall. I have written quite extensively of these years, mostly in Christmas letters to my children, a collection of which I anticipate publishing soon.

A Recollection of a Holiday Past.
Adapted from a letter of 1987.

It was great to be alive this Thanksgiving day of 1938. The manure was all cleaned from the barn, the cows had their hay and the hogs and chickens were fed. The wind had changed to the northwest and the snow was really coming down when I left the barn. I wished we had a radio, because the Packers were playing football in Detroit,

Ma reminded me that I had to go over to Mickey's and help with the milking that night, clean their barn and do the chores the next morning. I hurried through the storm because I wanted to get the cows fed before dark. I didn't mind milking with the kerosene lantern hung by the milk can, but throwing hay down from the mow to feed the cows was tough in the dark. Thank goodness, there were only twelve cows, and most of them were easy milkers.

The daydreams began while the milk pail slowly filled.

Maybe I would get to go to high school, as Art and Betty did, but college was out of the question with times so bad. But gee, wouldn't it be great to be an engineer? Design and build bridges and buildings and railroads and highways and ore docks -- the cows dirty tail hit me right across the back of the head. On to the next cow and next daydream. Maybe someday I would go out west like Pa did when he was young, riding the rods of the Soo Line freight and meeting all the hobos along the way. Guess the wheat fields are hard work but ---- I stripped out the last few squirts and emptied the milk into the strainer.

Only ten more to go and I could hear the blizzard in full fury whipping the snow across the yard and around the barn, and hoped it would let up for the walk home tomorrow.
jimmyj 1987

Our family slowly climbed out of poverty. Pa had fairly steady work for the county highway department. I began working away from home where ever I could, doing farm chores or making hay. I remember how proud I was to be able to buy a bicycle at the end of one summer. I never regretted learning to work to support myself. It was an interesting time, and we enjoyed inexpensive entertainment such as provided by the river, which prompted this recollection.

The Norwegian Bridge
It would be great if we could just drop back to 1941. I was twelve years old with time to walk the mile to the Black River at Norwegian Bridge.

Today's kids have swimming pools, gymnasiums, tennis courts and hockey rinks, but I had a river running slowly through a pasture -- oops, be careful where you step.

Here we would fish for suckers in spring. My mother pickled them -- we never caught very many. We fished with cane poles and a string and used angleworms for bait. We'd build a big bonfire on the bank, fish into the night, and walk home in the dark.

In the summer the kids from miles around would swim here, because it was the deepest pool we had, twelve feet or so, where water swirled around the piers that supported the two spans. The steel girders were iron riveted together, stood eight feet high, with a wooden plank deck in between to provide one-way traffic for cars. What a beauty of a bridge!

If you were careful, you could climb through the guard rail and stand on the heavy iron plate that covered the four-foot diameter iron pier, and dive into the water ten or twelve feet below. That was the test, to see if you were big enough, and could swim well enough, and had the guts to be one of the "big guys."

We knew some northern pike and a bass or two hid among the weeds along the shoreline. We never had much luck catching them.

One time I'd netted minnows from another little creek and threw them in the cow tank to keep for a different fishing trip. I borrowed my dad's casting rod and an old spooled casting reel, put some minnows in a bucket in the basket of my bike, and peddled to Norwegian Bridge.

The only way to fish the deepest hole was to sit on the pier, place a minnow on the hook and let it swim freely on the line six feet beneath the bobber. I had barely settled in, watching intently as the minnow pulled the bobber about, when I heard footsteps grating upon the planking of the bridge.

It was Otto Nelson, eighty and some years of age, leaning on a hand-made cane, walking down the bridge that bore his name. The Norwegian of the bridge, he had homesteaded here when a young man and raised a family. I think the youngest, Gordon, was still at home and had a little dairy just like many others in that area. He hobbled to the rail in back of me as I sat fishing off the pier.

"How's the fishin', son?" he asked. His accent was strong, but I was used to accents and understood him well.

"I haven't had a bite, Mr. Nelson," I replied.

He leaned over the rail and squinted down at me, "Oh, it's Chimmy, Louie's boy," he said. "Fishin' ain't much now days, but ven dey first put in da bridge, it vass diff'rent den. Diss fishin hole vass bigger den. Dere vass a muskie here, fifty inches long. He broke effry kint of tackle dat I had. I sed, I'll fix you, you darn son uff a gun.

"I cut a cedar block off of a big olt fence post, about fourteen inches long an' drillt a hole trew it. Den I took a closse line, wit da biggest hook dey made tied on.

"I hat a sucker, sixteen inches long at least, for bait. I figgered he woot take da bait an' wear himself out pulling on da cedar block. Den I woot come an' snag da line an' bring him in."

Very seriously, he continued, "I brot da whole darn verks right here up on da bridge, an' trew it in da vater, right where your bobber is."

He paused a moment, to let that sink in.

"You know what happened den?" he asked.

"I saw him come right up, and take da cedar block!" I nearly fell off of the pier.

I never caught a fish that day, I seldom did. But I never will forget old Otto's tale of how it used to be, when fish were bigger, smarter, and stronger than the men who fished Norwegian Bridge.

jimmyj 3/11/98

I was in the eighth grade when the Japanese attacked Pearl Harbor.

It was Monday morning, cold as hell, December 8, 1941. It was dark when I finished milking the cows and had shoveled out the manure, fed the old sow and the chickens. I had to carry a couple of five-gallon cans of water from the old cast iron pump in the yard for the cows to drink during the day.

Everything seemed to be strangely quiet, almost as if in shock. I latched the barn door and walked over to the house and into the kitchen. Ma had made some oatmeal for breakfast, and as I washed and sat down, she listened intently to a scratchy radio station WTMJ, in Milwaukee and the news of yesterday's bombing of Pearl Harbor by the Japanese.

"President Roosevelt is going to declare War," she said, as if we really understood what that meant. "The news is not good and they don't seem to know exactly what happened."

"Do you know where Pearl Harbor is, Jim?" she asked.

"I'm not sure." I said, "but I think it is the big naval base near Honolulu, in the Hawaiian Islands. Isn't Merlin in the Navy out there somewhere?"

"Oh, God," she said, "I hope not. Your Aunt Lucy has enough trouble as it is." As a matter of fact, he was there and did survive the bombing, then assisted in the recovery of the many casualties.

I had just turned thirteen and was in the eighth grade in the Little White School. It was about a mile and three quarters away and Marianne, who was in the fifth grade, and I had to walk to school. The Kovatch kids, the Kresses and the Adamzuks would often walk with us.

I recall Miss Huber didn't start classes right away as she carefully pointed out to everyone where we lived in relation to Hawaii, and where Japan was, and what it meant to declare war. It was a one-room school; so all eight grades listened intently. I had five classmates, and we were one of the bigger classes.

I remember the intense sense of patriotism everyone felt, and we kids wished we were old enough to "sign up." Almost everyone who was not tied down with a family and was of "fighting age" did. I remember how proud the Adamzuk kids were of their older brother, Russel, who was in the Marines. The Draft, which had been organized previously, took many more men, although deferments for farm work were sometimes available, depending on the local draft board's ability to make its "quota."

Many foodstuffs, gasoline, tires and shoes were soon to be rationed. To us, rationing was more of a bother than anything. Nobody in our neighborhood had any money for luxuries anyway. I remember Mom being concerned about getting enough sugar to do her canning.

Gasoline was rationed, but I no longer remember the specifics. I think an ordinary "A" coupon was good for five gallons a month, which would cost about a dollar. It seems we more often had the coupon than the dollar.

I remember going around the neighborhood collecting scrap iron and paper for the war effort. Work was available, even for kids, because there were few men around.

My folks would listen very attentively to the ominous voice of H. V. Kaltenborn, and the somewhat more subdued Lowel Thomas, reporting the war news. I also remember having to learn to identify the silhouettes of warplanes, and what a worship we had for pilots and other war heroes.

As I grew a little older there were a lot of girls around and I had Art's old '36 Chevy, a little farm gas, and it seemed, always a girlfriend. Then when the war was over, the GIs came home, took our girlfriends away and married them. Listening to the radio, I learned to love the big bands and the swing music, that we now call classical jazz.

It took many years for me to understand the emotion with which my folks and most grownups viewed the war. They had lived through the First World War and most of their parents had come to America to escape the senseless wars of Europe, and the conscription of young men into the service of an emperor or king. My folks viewed President Roosevelt almost as a god, not some mortal who also made mistakes.

We all adjusted to the realities of war, and before I was old enough to enter service, it was over. But I have never outgrown the impact it had on my life.

jimmyj 3/1/98

I guess the process of growing up was accelerated by the War. Both good habits, such as early rising and working hard, and bad habits, such as drinking beer came very early. Even the most spoiled kids from "in town" worked in summer, many at the pea cannery. Pa decided to grow green beans one summer, but as I remember, it was all work and not many beans.

I wonder how different it would be today if things had been a little better back in '44. My mother kept a journal, jotting down the status of our family during that year. My memory will add some things my mother was too proud to tell. She never would admit that we were poor, perhaps because we were so rich in family love and things that really count.

Louie, my Dad, or Pa we called him, lost his thumb and index finger in '34. He caught them in a silo filler that was used to chop the corn. A one handed man can't milk a cow; in '38 he lost the farm.

No insurance, and no job, he took all he salvaged with an auction, about eighteen hundred bucks, and bought a house on seven acres. We had cows, just two or three, an old sow, some chickens and the garden from which Mother canned the jars of winter food.

Then in 1940 he smashed the remaining fingers on that hand while repairing a rock crusher. By Nineteen-Forty-four Pa had fairly steady work, because all the healthy men were off to war.

Brother Art was in the Army, in Iran, trucking guns up to the Russians. Betty's Bill was in the Army, with the troops in Germany. Lucille's Gerry was in California in some kind of motor pool.

Brother Mickey's was trying to make a living on Grandpa Ludwig's original homestead when the farmhouse burned in '43. He'd found a job running a dairy for Mr. Fries in Kingston, Illinois, and wrote to Ma and asked if Jimmy might come down and work with him on the farm. He was milking forty Guernsey and had about three hundred acres under plow. Help was hard to find. Of course I could help out, but I had to find a way to get there, because it was about three hundred miles away.

Pa stopped in and talked to Mr. Nuendorf, who was trucking butter and dried milk from Medford to Chicago where they shipped it out to help the war. He arranged for me to ride a freight truck to the south side of Chicago at the docks. I was fifteen, and had never seen a city bigger than Wausau, Wisconsin except in picture books.

Pa drove me to Nuendorf's. I had a little cardboard suitcase; with a change or two of clothes and a lunch my mother had packed. I remember I had only two dollars. Somewhere we found a map of Illinois that showed some of Chicago on a somewhat larger scale.

We rode all night, and at dawn pulled into the freight docks. I showed the driver on my map where I was going. He directed me to walk out through the gate and catch a streetcar, one that headed west, and stay on it as far as it would go.

I'd never seen a streetcar, bought a pass for just two bits, but didn't know it was a transfer stub that I could use on other routes. I tried to track my progress on my map, but soon was lost. I rode until the streetcar stopped, the driver rushed me off, ignored my plea for help and left me standing on a corner on the west side of Chicago, God knows where.

I thought I'd ask directions in a bar as if I was in Stetsonville, but they just told me to get out. I asked a policeman -- he just laughed and shook his head. I walked until I found a street whose name I located on the map, and decided to go north to Des Plaines because it appeared to be a suburb where a Greyhound bus might stop.

I bought another ticket on a streetcar, sat down right behind the driver and asked him to not let me miss the street that took me to Des Plaines. He told me how to use my ticket for a transfer. It was mid-morning when I walked into a Greyhound bus stop.

I asked the lady for a ticket to Genoa, Illinois. It cost a dollar sixty-two, I had a dollar and a half. The lady looked at me and smiled and took twelve pennies from a change jar on the counter. I finished the last sandwich that my mother made for me, and then caught the bus after noon.

The map showed about fifty miles to ride. The bus left from the station and headed north on Highway 12. That wasn't where I meant to go at all. I asked the driver who said we were going to Genoa City, Wisconsin, just as my ticket said. I asked the driver to let me off so I could hitchhike west.

"I can't do that," he said, "We'll go on and I will put you on a bus that goes to Genoa, Illinois. I'll tell them it was our mistake and we will make it right." He put me on the proper bus; I thanked him from my heart.

It was after five when I arrived in Genoa, I walked out to the edge of town to hitch a ride. A very nice old farmer stopped his rattletrap pickup and took me the five miles to Kingston, where he had to turn off. I walked the last half-mile from Kingston on the highway to the farm. I went over to the milk house where milk was running on the cooler. I filled a cup and drank it almost in one gulp.

Mickey and Bob Schoonover were just about to finish milking for the night. "How was the trip?" they asked. "We thought you'd be here before lunch."

"I had a little trouble. I am not a city boy. I'll tell you all about it while we eat, right now I'm starved."

The work was hard and the days were long. Seven days a week, from dawn to dusk, except for Sunday afternoon. Three men had all that they could handle. Bob soon left, off to the service, and Henry Zielke came to work.

Work on a dairy in those years was pretty much routine. Before the dawn, a young guy (mostly me) would get the cows in from the field and bring them to the barn. Each cow had a specific stall and found it easily. The ground-up grain they ate was scooped from a cart.

The milk machines were sterilized and the udder was scrubbed clean. The milker hung beneath the cow on body straps and vacuum tit cups were attached to milk her dry.

The milk was carried to the cooler in the milk house, then put it in ten-gallon cans and kept cool until a truck would pick it up to process for delivery in Chicago the next day. That was the modern way then. Today it's really changed.

With the milking finished, we turned out the cows and cleaned the barn, then went over to the house for breakfast.

After breakfast Mickey would assign work. We worked until noon, then took a break to eat, worked until five, then milked the cows again. When we were finished, and had eaten supper, we could take a break before we went to bed. Sleep came easily.

Some days I'd ride the tractor, cultivating corn from dawn to dusk. We had never heard of herbicides.

Putting up alfalfa hay was about the hardest chore. We pulled a mower with a little Case tractor. When the hay had dried, we pulled a side delivery rake to make the windrows for the baler. A man would ride each side of a New Holland baler, moving up the bale dividers, insert wire and make a knot. The dust would be so thick you could hardly see, but you dare not miss a lick, for if you did the rig would have to stop.

Bucking those alfalfa bales, weighing 60-100 pounds, sure made a man of me. Before I quit for school, I could hoist a bale over my head and throw it on a stack. We had no bale pickers, just loaded them by hand on a wagon, and then stacked them in the barn.

We had a horse, an old three-gaited mare that we used to get the cows because we had to cross a river and drive them through a muddy underpass beneath the tracks. One day she dumped me in the river when I was standing in the saddle, fixing fence. I have disliked horses ever since.

I met a lot of local boys and played baseball in a 17 and under league. We all were from the farms, practiced twice a week after work was done. We played a game on Sunday afternoon.

The next year I worked for Uncle Oscar, my mother's brother, who always listened to the news and tried his very best to make me think. He sparked the hope for college, something I had written off as unaffordable. I wanted to be an engineer, not kill myself while working on a farm.

Mom ended her journal with these words in 1944, and let me quote, "Jim worked for Harold and earned almost one hundred dollars. Things are pretty much OK."

High school was a dream come true. I loved it. The school bus would pick me up about seven, so I had to do the chores before then. I went out for football, had to walk home seven miles after practice unless I could hitch hike a ride, and then had the farm chores to do after I got home. Pa thought I was crazy to play football, but it was a real confidence builder for me. I realized that even though I couldn't see well, I was just as tough and smart as anyone.

The Coach in '96

As class reunions go, it wasn't much. Not that the old friends I knew did not try hard. They worked; I didn't, so I'll not complain. But after fifty years, there was only one emotional moment I won't forget.

I looked up from my drink and through the crowd, saw someone push a wheelchair through the door. My God, that's Coach, I thought, Glen Derouin, and I thought he was dead.

Johnny said, "Oh no, he's eighty some and lost his legs from diabetes years ago. He'll be glad to see you, he's asked about you now and then."

The crowd had gathered 'round him, so I waited for a chance when it had cleared somewhat. A lump formed in my throat, a tear welled in my eye as I extended my hand. Here was a man who shaped my life more than anyone other than my parents.

He grasped my hand in both of his, "Jim?"

"It's me, Glen, how are you?"

"I'm just fine, Jim. Shorter than I used to be and can't catch passes quite as well. Do you remember how we played that game at Chippewa Falls? I've had other undefeated teams in forty years in high schools in this state. But you guys were my favorites."

Then he went on to tell me about a game that had seared into my mind some fifty years before.

September 1945, cold and dreary but America was feeling very well. We had won the war in Europe and now had to finish off Japan.

Friday afternoon and we had come to Chippewa Falls to play a football game against Coach Derouin's alma mater. They were a big high school with five times as many students as we had. I think Coach thought that it might be good to have our cocky ears pinned back a bit.

As we gathered at the sideline, Coach extended his hand, to be covered by each hand of ours in turn. There was no sound, as Coach looked each of us right in the eye, and then in calm and solemn tone, he simply said,

"We can win!"

I was center on that team, and captain too. I was little at 149, and each of us played both offence and defense in those days. On defense I would back the line on the left side.

Tommy ran the kickoff back to the thirty-three yard line and we lined up to play, first down and ten. As I bent down to center, I looked straight into the eyes of 220 pounds of beef, the defensive tackle in front of me. With a sneer and smirk he cackled, "So what makes you farmers think that you can play a game with us?"

I stood straight up and formed my hands into a T. The whistle blew, time out. Coach was waving from the sidelines, wondering what in hell was going on. We went back to a huddle and I said, "Now listen up! Do you know what that bastard said to me?" I told them what the wise guy said, and then, "Let's show him how we country kids play ball."

We had fire in our eye when we went back to the line. I gave the ball to John, who handed tight inside to Beaner. Emil Joe and Ray had put the wise guy on his back. I slipped on through and took the linebacker out and Beaner ran for twenty yards in the hole.

Then we lined up and did it all again. It was 20-0 at halftime; we held on for a win.

How often in my lifetime has that lesson pulled me through? If you have a real team, there is no game that an emotional high and confidence can't win.

When Coach finished with the story, he still held my hand in his. The fire in those steely eyes had not dimmed, not a bit. I tried to tell him how his way had helped shape my life. I couldn't speak. A lump came in my throat and tears filled my eyes.

I pulled my hand from his and walked away. I would never ever, ever let Coach Derouin see me cry.

jimmyj 2/3/97

High School was easy for me, so I never learned good study habits and goofed off a lot. I discovered girls and always seemed to have a special one. From somewhere I developed a rebellious streak, which I have never lost. In our senior year, I was again the class president and helped to orchestrate a senior skip day, which was unheard of at that time. We felt that money we had earned through sponsoring school activities and a lot of hard work should be used to bring in a really popular big band for our senior ball. The administration refused, and spent the money on a school-wide public address system. This is how the skip day came about.

It was Friday morning, early spring of 1946. The senior class had gathered for a meeting, 110 strong, in the main assembly room of Medford High. As I walked to the podium I noticed that our faculty class advisor wasn't there. That was a definite call for anarchy.

"Will the meeting come to order," I intoned. "May I hear minutes from our previous meeting?"

They were read and approved, old business quickly done.

"Is there any other business we should do?"

A hand was raised. I said, "I recognize the gentleman in back."

"I think we should have a senior skip day just for fun," he said.

"Well, clarify yourself and put it in a motion for a vote," I replied. He promptly did, and then a second was obtained.

"Will all in favor please respond by saying 'Aye,' " I said.

An "Aye!" resounded through the room.

"And those against say, 'No'."

A shuffling of feet, a couple timid voices answered, "No."

"The Ayes have it. Someone make a motion for a date and time," I said.

Someone shouted, "The first nice day next week!"

"Sounds good," said another. "Bring a car if you can!"

It was obvious the meeting was descending to a mob. "I will accept a motion to adjourn," I loudly called, and that was duly done.

Some of the students left, the others gathered 'round. "How will we know for sure the skip day's on?"

"I am assigned to this room during the first period. I will stand up and announce a senior skip day," I said. "Then we will go around and call the others out of class, and tell them to gather at the western door. You guys with cars bring them around and we will split up as we wish."

The first nice day was the following Monday, and the plan worked like clockwork.

As punishment for our misbehavior, everyone who participated received a blank paper at the usual graduation exercises. We had to return to school for three punishment days in order to receive a signed diploma.

jimmyj 9/01/96

Chapter 4, Off to College 1946

My dream of being an engineer looked pretty bleak in 1946. It was the one time in my life I felt a sense of hopelessness, even despair.

Many of my friends were going off to college. We had finished high school that spring, care free, full of hope. We were working, playing baseball on Sundays and drinking too much beer.

When the summer ended I had hoped to go to Stevens Point to the state college. When I arrived to register, the school was overwhelmed with fellows out of service on the GI bill. The classes that I needed for an engineering degree were full. It would have been ridiculous to waste my hard-earned money, and so I returned home. I went back to driving truck, but that work ended in the fall. I got sixty cents an hour for plowing snow, but work was so sporadic I could save little.

At Christmas time my friends came home from school and we had the usual parties. I felt out of place and got advice by the bushel, sympathy, and little else.

After Christmas, I was really in the dumps. When the war had ended the draft was discontinued. The GI bill was over, but I thought that I should join the Army anyway. Mom and Art, who had returned from service, convinced me not to even think of that. It was a little tough to keep my head up and my hope was almost gone.

I've always been a lucky guy. When things were really down, that's when an unexpected piece of luck occurred. Pa called from the county shop. A surveyor had come from the State engineering office in Eau Claire. They needed to get data from a survey on County Trunk M, which was gravel at that time. He needed someone to assist him, holding a target rod and driving marking pins. He asked if I would be interested. I didn't hesitate a bit and told him I would be right there.

Then I met Carl Wojan, and what a friend "Wojo" turned out to be. Carl was as common as a cotton shirt, reminded me of Icabod of Sleepy Hollow fame. Tall, gangly, shuffled about, talked as if his mouth was full of marbles with a caramel thrown in.

He had spent six years in the Navy and fought a lot of the war for Uncle Sam, but wouldn't talk about it. He was working as a surveyor until he could get back into the class cycle at Wisconsin Tech after being discharged.

He came from Eau Claire with a state truck and a set of specs to do the survey. His gear was thrown in back, with almost everything he owned stuffed into a duffel bag. The most amazing thing to me was that wherever he set his foot, he was at home. He was my first acquaintance with a real engineer.

We spent two weeks together, all of the short and cold daylight hours out on the road. He introduced me to surveying. While we warmed up in the truck, or traveled out to work, we talked about his engineering school. He told me how he'd worked his way through the two years before the Navy called, and said he saw no reason why I couldn't also do that.

He was anxious to return to school. With his GI income, he wouldn't have to work all of the time. He told me where to write and what to ask for, who to talk to about a job, and where to cheaply room and board.

His enthusiasm was contagious. I vowed that when the autumn came I'd be in college.

I learned that registration was to be simple. I needed to arrive with a little cash and my high school transcript on the day assigned. They thought they would have room for everyone; the two previous years had taken off the pressure of the GIs wishing to enroll.

Two things happened that next summer that are indelible in my memory.

One evening on the way home from work, Pa and I stopped at John Doctor's for a beer. I told Pa I had just completed making all arrangements to go to college. I asked him, if I needed, could he help me with some money.

He looked down at his beer and thought a moment, then slowly said, "Jim, maybe things are better than they used to be; we are not so hopelessly in debt. But I have no reason to think you are more deserving than your older brothers were. I couldn't help them so I won't help you. It would not be fair.

"Maybe if you really need a little cash sometime, I might lend you some." He set down his beer and looked me squarely in the eye, "The interest rate is six percent, and by God, I want it back!"

I've thought about that conversation many times, and have told this tale to some of you. It was one of the best things that ever happened to me. I worked my way through college and never had to get a nickel from my dad.

The second memorable event found me walking down the street in Stetsonville one day. As I passed by the bank, Art Greiner called to me.

"Come in a minute, Jim," he said, "I hear you're going off to school."

"I sure hope so, Mr. Greiner," I replied.

We walked into his office and sat down. "Well, tell me all about it, Jim."

So I explained how much I hoped to be an engineer and was going to Platteville in the fall." He listened quite intently, and then he reached in a drawer, pulled out a blank checkbook and handed it to me. "Jim, there may come a time when all the chips are gone and you will wonder where to turn. Write a check for what you need; have them call me if they wish. When the check gets here we will take care of it some way. I'll call your dad and we'll set up a loan or some account, but don't you worry, it will be OK."

I carried that blank checkbook for three years; never used it once.

Art has long been gone, and so is Pa. I never will forget the faith they had in me. I don't think I've disappointed them.

I took the Greyhound bus south to Platteville in the fall. Everything I owned was in a cardboard box. Wojo helped me find a room. I put my savings in a bank and I was ready to become a Civil Engineer.

The next morning I reported to the school. We sat in a line that slowly moved down the hall. The GIs knew how to cool their heels in line, but I was learning something new.

When my turn came, I handed a copy of my high school transcript to Dale Dixon, one of the professors. He looked at my transcript carefully. He said, "Anyone with grades like this should really be a Mining Engineer. I see that you signed up for Civil. Why don't you change?"

"I don't know," I said. "What is the difference anyway?"

"We'll teach you everything that Civil does," he said. "Then on top of that, all about geology and mineralogy. There is a lot more opportunity to really make some bucks in mining. Why don't you give it a shot?"

I thought a minute. I was paying my own way -- why not get the best I could?

Just like that, I decided to become a Mining Engineer, and signed up on the spot.

It was a choice I never have regretted through all the years.

jimmyj 2/20/98

Here are some stories of my days in college. I really enjoyed the common and friendly professors and the relaxed attitude. Yet there was a sense of urgency as the ex-servicemen hurried to get on with their lives.

Check your Bubbles, Boys

Nowadays it has a high fallutin name, University of Wisconsin-Platteville, but in my time we called it "Whiskey Tech."

Back in 1853, the good citizens of Platteville had taken native limestone and built an academy, now known as Roundtree Hall. In July of 1907 the State Legislature passed a bill to establish a Wisconsin Mining Trade School. This building then became The Wisconsin Mining School.

In the thirties it was changed again to The Wisconsin Institute of Technology.

When I started in 1947, there were about three hundred guys, mostly ex-GIs, a couple of girls, and a rather little group of excellent Profs. "Uncle" Milton Melcher was the chancellor and ran the whole school with the help of an office staff.

Orth and Ottensman taught math. Reverend Doering was the English Professor. The other professors were Pinky Harker, Proffy Pett, Willie Broughton, Major Kurtz, Dale Dixon, Jughead Clarke, and Dobby Dobson. We them all and they knew all of us.

Old "Dobby" Dobson had taught chemistry since that first class in 1907. Honestly! He was a friend of that Chicago great, Amos Alonzo Stagg, and had coached the football team ever since those early days. I played for him one year; he was in his eighties then.

He never learned my name. He called me "Number Ten," the number on my practice jersey, no matter what I wore, if I was in class or in a game. His playbook from the twenties was known by heart by every coach we played.

We set up in a single wing; the quarterback a blocking back, the tailback was a drifter, he or the flanker were in motion every play. My job as center was to hit that motion guy with the pass as he moved by.

I had to look back upside down to see just where to aim the ball. Then the guard would pull, he and the other backs would lead the ball into the hole. I got clobbered every play. It was a mess and we never won a game. We might have if some of the ex-GIs had shown up sober once or twice. At least it was a game, not like professional mayhem that you see today, but mayhem nonetheless.

One day I threw a block to stop a runner on the sideline, rolled over him into the crowd and flattened Jughead Clarke, whose son was playing in the game. I broke his arm. He wouldn't leave until the game was over, then he went to have it set.

Jughead taught surveying. He was a little wizened guy, bald headed, very short, about five foot four. His voice was high and raspy, I guess squeaky is the word.

In the spring the class would go out to the golf course to survey -- a laboratory class the book said, a picnic in our minds. We would go to school, check out a transit with a rod, some tape and pins. We each kept a set of notes in our own survey book, which would determine our grade.

On the way from Roundtree to the Platteville Country Club, you had to go up Two Street, right by all the saloons. So most of us would stop and get a six-pack just in case our thirst would get the best of us as we honed our skills. There were a couple crews of ex-GIs that always drank a round or two. I remember that Joe Gambil and Joe Becker always kept their books under a hat rack at Earl's Bar.

Jughead had us all dispersed around the eighty acres, and of course, we spread as far as possible, so he had to walk a long, long way. When he would amble up, the first thing that he'd say was, "Check your bubbles, boys." Then he would elbow you aside, look in the eyepiece of the transit, check to see if you were set up right, and then ask to see your book. As he would leave he'd say it one more time, "Don't forget to check your bubbles, boys."

One day Savoldelli, Gambil, Becker and a kid named "Hooley" Plourde stopped at Earl's Bar and quickly guzzled several rounds. When they rushed off to practice surveying, they left part of their gear along the bar.

Jughead checked them in with one cocked eye; he could see they'd had a few. They hurried off to set the transit. Then they realized they didn't have the tape or pins. So they decided Savoldelli, who was six feet four, should set the transit and set it high. The eyepiece was a good six feet above the ground. Then they began to make believe they really were surveying.

Soon they heard the old familiar call, "Check your bubbles, boys," which Savoldelli carefully did. Joe had been sitting on the ground, making up some survey notes, judiciously recording in the book. Savoldelli waved his arms at Hooley in the distance, as Becker knelt as if to drive a pin. When Jug walked up he saw the eyepiece way up there. He couldn't reach it even standing on his tip-toes. He pointedly ignored the transit as Joe handed him the book. He just glanced at it, and with a cheery, "Check your bubbles, boys," was on his way. As he walked off they opened up another beer and when he was out of sight over the hill, they picked up everything and went back to the bar.

We laughed about it then, made fun of the old man, and I still chuckle when I now recall. What a place to learn to be a mining engineer.

I'll tell you one thing, though, I never touch a transit or a level without hearing from the past, hearing that high and squeaky voice say, "Check your bubbles, boys," which then I faithfully will do.

My old friend, Mr. Clarke, knew exactly how to teach a bunch of crazy guys.

jimmyj 2/1/98

I thoroughly enjoyed college. It was amazing how little a man could live on when he didn't have any money. I held some crazy jobs now and then. The first year I washed windows for the neighborhood on Saturdays, and then got a job setting pins at the bowling alley. About early March, Forest Hanson and I took over the job of greens men at the country club. Forest was able to live in the caretaker's house for free, so he stayed on through the summer, and then I came back in the fall. I worked there each of my first three years.

One winter I got a job as night man in a chick hatchery. Every two hours, I had to manually tip the egg trays in the incubators, and then register in the time clock so they knew it had been done. The only good thing about it was I could use a desk and a calculator to do my homework. I quit when the frost left the ground at the golf course. I learned to wake up without an alarm, and drop right back to sleep. Don't use an alarm clock to this day.

The teachers' college, then Wisconsin State at Platteville, was just down Main Street and there was such an intense rivalry that the sports teams could not play one another. They called us "miners" and we called them "tits." It was the bounding duty of every miner to figure out a way to light the teacher's homecoming bonfire ahead of time, or some other nonsense, such as replacing the ticket booth at the football field with an outdoor toilet, or dropping a charge of dynamite down the goal post.

College stories are a dime a dozen, boys will be boys, have fun and chase girls, unless you're broke.

In the spring of '49, before the last class at Whiskey Tech was complete, John Thompson, Sonny Sutherland and I lined up a bachelors pad to live in when the school resumed that fall. It was located in downtown Platteville, on the second floor above a grocery store. It was an old building, with high ceilings and hardwood floors, which had seen a million mops or more.

Cast iron radiators clanged and banged, as an ancient furnace pushed steam through gurgling pipes. The tall windows were supposed to lift to let in some air, but some would not move at all. A window in the transom up above the door would sometimes open, sometimes not, and when it did the feet of children banged like gunfire on the hardwood in the hall.

An open double burner gas stove was on one wall, next to a cast iron sink, with no drain board. We bought a little table at a junk store, put an old wastebasket under it, set that along the wall next to a cupboard from the civil war. It had curtains made of gingham and no doors.

Some tenant long ago had left a table and some chairs. We bought three army surplus, wood framed, canvas folding cots. Two were in one room, one in the other for John, the senior member of our crew.

We called it home.

John and Sonny were from Mercer, Wisconsin, another hundred miles north of my hometown of Stetsonville, which was two hundred miles away. Both had been sailors in the war and were at Whiskey Tech thanks to the GI bill.

Sonny was a tough little guy, and a boxer. He had been in the thick of the Pacific war. He'd have a drink or two and wartime memories would shake him up for hours. Sometimes he'd drink some more as if to blot that part of life away, and more than once we were called to a second street saloon to bring him home. Most of the time he was all smiles and fun, but he had to work hard. The engineering classes weren't for fools.

John was quiet, curly-haired and sort of slight. He had lost his ship and nearly all the crew of a sub tender to a Japanese torpedo during the war. He survived because he was on deck, guarding moonshine being made of prunes from other equally thirsty sailors. After he was picked up by the sub that sank them, he spent a little time in prison camp, but thank God the war was nearly over. He was almost like my brother and went to Climax with me the next year.

Those sailors organized the household very well. We would each put a little cash into a grocery pool, and one guy would do all cooking for the day. The other two did up the dishes and swept the floor.

Saturday mornings we swabbed the deck, washed our clothes and shopped for groceries. My mother sent down canned goods from her garden and canned liver sausage from a home-butchered pig.

John supplied canned venison, and Sonny contributed smoked and frozen fish. Eggs were about fifteen cents a dozen and were the breakfast staple. We rented a locker in a freezer locker plant, because we had no refrigerator.

When rabbit hunting season came, we took a bunch of guys to hunt the cornfields for the cottontails that fattened on the corn. We filled the locker, and had enough left over to feed all our friends that night at a party.

Admission was a six-pack, any kind.

When John came back from Christmas vacation, he brought some blocks of frozen herring and a stone crock. We went to the Silver Dollar Saloon, just across from Ken's Bar, and the bartender went to the basement and brought some cheap old port wine for us to be used for pickling. They tasted really great, except as each of our friends walked by the crock he'd have to try one, so when the fish were pickled well, they were nearly gone. The five-gallon crock was refilled three times.

Our friends often stopped in and did homework on the kitchen table into the night, then would go to Second Street to the saloons, which were just a block away.

We frequently used a pressure cooker to prepare meat. We would dump in a quart of venison all cut in chunks, some onions and some garlic, salt and pepper to taste. Flour with some water was mixed in a jar and was added before turning on the gas. Potatoes were boiled on the stove, vegetables were heated and you were ready to set the table.

We would thaw some rabbit from the locker plant, brown it well and cook it in the pressure cooker. You know, I still love meat that's cooked that way.

As I think back, that time was most enjoyable. We were always broke. I had to work a lot -- we didn't party much.

Most GIs did their drinking straight up at the bar, but when they went to school, it was a serious business. They were determined to make up for what they'd missed. They spawned the baby boomers that rule the roost today.

Many of them sit around these days and contemplate the past. School is remembered but the memories of the war are denied. College was the best thing they had ever done, and batching with them at Whiskey Tech was great fun. I returned to Platteville to recruit Engineers for Climax several times in the mid sixties. The school no longer offers a degree in Mine Engineering, nor does hardly anyone else. I guess I shouldn't care, but I do. I wonder where the poor kids go that need to work their way through school? Do they ever look back from day labor jobs, and curse their luck? Where will the expertise come from to move the earth that moves the world?

jimmyj 3/9/98

Roundtree Hall

At the end of the school year John and I went to Climax and in the fall John went to the Missouri School of Mines at Rolla. I Stayed at Climax and returned to Platteville in the fall of 1951.

During the fall and winter of 1950 and through the summer of 1951 I was working as a contract miner at the Climax Mine. Somewhat to my surprise, I found that the mine and its people fascinated me. Even though I had acquired a wife and a brand new baby, we decided that we should return to complete the studies necessary to receive a degree. Here's a tale about those times.

An Easter Snowstorm

By the spring of 1951, I knew that mining was the life for me, but I knew there really was no future unless I earned a degree. So in the fall of '51 we bundled up our babe, took my mining contract earnings and headed back to Whiskey Tech. We moved into a tiny trailer house, a relic from the war that had housed GIs for training at the school. A community wash house, with some showers and some not-so-private toilets served our needs. When it got cold, the bedclothes behind the bed froze to the floor and a bathroom trip that winter was a chore.

And yet it really was a happy time -- new friends, no money, working when and wherever I could. I worked some as a greens man in the fall, then ran a little lead zinc mill each night in winter until we ran out of ore.

Schoolwork was fun for me; Climax became a point of reference that made my engineering studies easier. I breezed through the year.

When Easter came and graduation day was near, we decided to go and see my folks. Stetsonville, two hundred miles to the north was where I had been raised. My folks had left the farm, now lived in town; right by the Catholic Church, in what had been my Grandpa Ludwig's house. My sister, Marianne, and her husband Gene, lived upstairs in what had been my Grandma Peissig's room.

Holy Week in those days was a solemn time. Most people had to work. We fasted through Saturday and helped my mother clean the church. Quiet visits, relatives and friends all were subdued in mourning for His death.

But Easter Sunday all that changed, for He had risen. A lot of laughter, sumptuous meals, and then for all we young folks, an Easter dance, for those were dancing days and we couldn't wait.

Sometime on Saturday, silently, as if on owl's wings, the snow began. Before the Sunday dawn, eight inches lay upon the ground; the wind began to move it all around. Ominously the drifts rose.

Pa paced nervously, he worked for the county and it was his job to see that all the highways in the county remained open. By seven in the morning, he began to call for crews to go to the county shop in Medford, to pick up a plow and begin clearing the snow.

But it was Easter time and many men were gone. He was short on crews. He asked Gene and me to help him out -- he knew we both had plowed snow before. I needed the money, that was for sure, but we had set our mind on celebrating at an Easter dance. No matter, because he really needed us, so we agreed.

Pa assigned us to a Dodge with a side plow and a wing. We made a swing through all the county highways to the south and east, then back again. It seemed the storm had eased a bit at noon. We came back to the shop at 3 p.m.

As we descended from our truck the phone was ringing madly. It was the sheriff, who told us a lady was in labor in the town Deer Creek, and her farmhouse was snowed in. We got directions, jumped back in the plow and started south toward Stetsonville.

The snow was coming down hard again; the wind was whipping it around. The drifts were building as we started east on Highway A. We started to see cars stuck in the drifts and we had to help them out. It was extremely hard to see.

Out by Shorty Brunner's place, a car was stuck in the right lane, nearly buried in the snow. We thought the wing would clear when we passed to the left. As we cruised by, the wing tip nicked the car below the windows and cut it as if it was a knife. We stopped but there was no one around. We ran into Shorty's house and called the sheriff, then Pa, and told them what we'd done.

The sheriff said to continue and plow out the farmhouse, then stop on our return and he would meet us there. The storm was whipping wildly again.

We went a couple of miles further east, then on a township gravel road to the south into the farmhouse driveway. It was completely snowed in; there was no way a car could leave without some help.

That was one happy Dad who bundled up his wife into his car that was warm and ready. He followed us until we turned back west on A, then he waved, and, smiling broadly, swung around us, heading for the hospital, about ten miles away.

The sheriff was waiting for us at the damaged car. He slowly shook his head. We had no excuse, so he took all the necessary data and sent us on our way. I guess everything came out all right because I never heard another word about our careless accident.

The storm seemed to abate a bit as we pulled in the shop. We hustled back to Stetsonville. My mother had delayed the Easter dinner just for us, and what a feast it was! While she watched Debbie we two young couples headed out to celebrate the end of Lent.

We needed no excuse to party in those days. We went to Jake's and John's and Joe's, the three saloons in Stetsonville; I don't think we danced a dance all night; just visited, drank some beer, and tried to help Gene and his brothers sing. It was a happy time.

That was the only Easter that my wife and baby ever spent with Mom and Dad. In just a couple of years my mom was gone. I returned that June to Climax, once more working in the mine, and trips back home were few and far between.

I've seen a lot of Easters come and go and always go to Mass just like I should. -- except for the one year a wild and woolly snowstorm had me plowing snow. It will always come to mind when I look out to check the weather on an Easter morn.

jimmyj 1/17/98

Chapter 5, At Climax Again, 1952

The return to Climax was uneventful. Our family was able to step right back in where we left off. Work in the mine had a new sense of purpose as the uncertainty about a career ended and I had become a Mining Engineer. We stayed with my wife Mickey's folks until a duplex on Fifth Street was available for us. The camp was crowded and new house construction underway.

The duplex rent was twenty dollars a month. Our heat was from the main steam plant, but cooking was done on a coal stove. Since we had many friends from the previous year, it was easy to fit in. I began playing baseball again and since we had so little, we expected little. We often slipped off to Glenwood Springs to visit Mickey's sister and I recalled those days with this story.

Oscar and the Brothers-in-Law
Glenwood Springs was something else.

Pool and Colorado Hotel, 1905

Something else than Climax, where winter lasts until June, and then you have a little bit of summer until winter starts again. Now I didn't really mind the winter, there were so many things to do and contract mining wore you down a bit, and sleep came easily at night. But back in '51 I learned the meaning of the malady called cabin fever, when Mickey said, "Lets go! I don't care where, but out of here, before I lose my mind."

So we would go to Glenwood Springs to see Bud and Lorena. The pool was hot, the weather warm and green grass in the fields looked so much better than the snow. We went to Glenwood often. We were newly weds, they were in their forth year, and our families had just begun.

It was on one such occasion, the sisters so involved in conversation -- never gossip as only sisters do, that Bud suggested he and I should ride about the town, check it out, perhaps stop in the Lariat and have a beer or two.

Now sisters understand immediately that husbands are not necessarily up to any good when visiting the bars in any town, especially Bud and me in Glenwood Springs. So one of them suggested that Bud should really take me up the valley to Basalt, to show me where they used to live, where Mickey had been born, and perhaps stop to see Aunt Lena, and introduce me to their Uncle Oscar Blanc whom I had never met.

It was a great idea, I guess, though later it was viewed with some regret.

Glenwood in those days was just a quiet little town. Across the river to the north was the Hotel Colorado and very little else. A few blocks south of old Saint Stephens, it was country, the high school nearly out of town.

Highway 82 was just a country road, by standards of today, and the trip up to Basalt was just a look at ranches all along the way. We didn't take the time to wander into Carbondale; it was a little mining town over to the west across the Roaring Fork. We didn't go to greater downtown in Basalt, but took the little back road up the hill to Oscar's place.

As I recall, his place was one of those you only see in pictures in the "Country" magazines. The flowers and the gardens all reflected Lena's loving care. The house was bright and scrubbed spectacularly clean, demonstrating care that only lack of ready cash will bring.

Oscar was a rotund little man, friendly, jolly. He welcomed us profoundly as the lord of his domain. He showed us all about his place, recounting all his gardening success, then took us to the kitchen and inquired, "Perhaps you would enjoy a little glass of wine?"

Of course we would, we were guests after all.

Oscar reached up into the kitchen cupboard, got a pitcher and some water glasses, which he set before us on the table. Then he went out to the cellar and soon came with a pitcher full of wine.

"I hope you like my Dago Red. It's from two years ago, which was a great year for the grapes. We always have them ship some in, from Washington, I think. There are just enough of us to make it worth their while. I don't think Frank, your father-in-law makes his own anymore. You young folks don't appreciate how a good glass of wine with meals improves your health."

He filled the glasses to the top.

"So here's a toast to you young men, and welcome to our family again!"

We raised our glasses high, and then took a drink.

Now I'm a Dutchman's Dutchman, and no stranger to the sauce, but this stuff took my breath away. It was dry, or maybe bitter. Sour is the word. It was awful, that first taste.

I grit my teeth and tried to keep the tears out of my eyes. I took another sip. Bud and Oscar nodded in approval of the wine. I must have been missing something, so I took another swig. It really wasn't bad as I had thought and soon my glass was down an inch.

Oscar promptly filled it up again.

The conversation varied widely. Uncle Oscar soon knew all about the mine, my family history, Mickey's health and Grandpa's garden on the hill.

The wine tasted better now, and the glass was filled and filled again. Oscar went out to the cellar, refilled the pitcher, brought it in to drink with home made bread and cheese that Lena had provided out of nowhere that I could see. The Dago Red tasted very, very, very good.

My thongue was getting tick when we decided we must leave. I hadn't noticed how the kitchen floor would pitch and roll. I couldn't find my butt with both hands. The thanks and goodbyes were impassioned -- it was a visit I would not forget.

Bud decided that as long as we were there we might go up the Frying Pan a way, and visit other friends and places that he knew. There was a bar on Main Street -- it's the only place that I recall -- and the friends were lost in Dago Red.

It was well into the night when we returned to Glenwood Springs and rather meekly stumbled in the door. Two red head sisters waited there. I don't recall, thank goodness, the words they said to us, but they created an impression that remains.

Another year, another spring, another visit down to Glenwood Springs.

I suggested that a trip to Uncle Oscar's would be fun.

Lorena didn't bat an eye when she replied, "If you can't stand it here with us, why don't you just walk to the Lariat and have a beer or two?"

I heard her say something about "the better of two evils" as we went out the door.

jimmyj 9/20/97

Life in Climax

My second daughter, Linda was born while we lived on Fifth Street. Our garage was up behind the house, on sixth. In January it snowed every day and not knowing when nature would call, I carefully kept the snow shoveled from the garage doors to the street. It seemed the plow would pass further and further away, and every pass of the plow would fill my carefully shoveled path.

That is until the night we quickly loaded up to go to St. Vincent's in Leadville. That night as we pulled away, I did not stop to close the garage doors. When I returned, a proud father, the snowplow had come by very close to the garage, took the doors with it and filled my garage with hard packed snow. I bet the guys down at the transportation shop laughed for a week.

The company tried very hard to provide entertainment and promote a sense of community. In the summer softball league, and the company picnic. In the winter there would be movies at the Rec Hall, bowling and basketball leagues, and of course the Sadie Hawkins Dance patterned after the cartoon "Lil Abner." Everyone wore a costume of some kind, the music was swing time from the war, and booze flowed freely if not openly between friends, mostly while stepping out the door into the cold.

Luke Weigang invited me out for a drink. We walked down the road a bit and he stepped out into an apparently empty, knee-deep snowfield, reached down and picked up a pint bottle of Wild Turkey. "How did you know it was there?" I asked.

"Have another sip and I'll show you," he said.

He screwed the cap on tight and tossed it fifteen feet into a smooth unscathed area. "See," he said. "It just leaves a little hole and the high school kids will never find that."

"But what if you forget where it is?" I asked.

"If I can't find it, I don't need another drink anyway."

Writing of Luke Weigang reminds me of the boats we built in 1950.

During that winter, while I was working as a contract miner, invariably lunchroom talk would turn to hunting, fishing and other pursuits we hoped to do when winter was over.

Bob Prior, Harry Barnes and Doc Murphy decided it would be great fun to build a boat for use on Twin Lakes to troll for trout. In fact, why not two boats? They asked me if I would like to help in trade for a little fishing time the next summer, and I agreed.

There was great controversy over the kind of boat we were capable of building, considering the cost of material, building space and time to get it ready for spring.

It was decided that we could build two fifteen foot long cabin cruisers, (it always rained on Twin Lakes), with a plywood hull covered with fiberglass, which in those days was a rather new material. It would be powered by a relatively small outboard motor. Water skiing was unknown, and trolling was at a slow speed. Bob ordered the plans.

All the technical considerations really meant very little because the real intended use was to sit on a beautiful lake, reasonably sheltered and drink beer. In fact, the usual invitation was, "Why don't you bring a case of beer down to Twin Lakes and we'll go fishing." In addition, the beer consumed while building the boats probably cost more than all the material combined.

It became quite a project. Each of us had been a concrete crew form carpenter at one time, but that is a long way from building boats. We had a lot of free advice from friends who just happened to drop by for a beer.

The crafts performed reasonably well. I do not remember catching a fish, although I do remember rescuing some people who had the poor judgment to get on the lake in a sailboat just before one of those vicious summer thunderstorms came over Independence Pass.

One afternoon Luke Weigang went fishing with Harry and the following conversation took place.

Luke said, "I sure wish I owned a boat like this."

"Tell you what, Luke," Harry replied. "You can have my share of this one if you will just furnish the beer for the rest of the summer."

"Wow, really! That's a deal." He and Harry solemnly shook hands.

The next day Luke sobered up and calculated how much that beer could cost, then spent a week convincing Harry to forget about it.

I didn't fish often because I returned to playing baseball for Climax, but those boats were around for quite a few years.

jimmyj 2/10/05

Writing a book about Climax

In a short twenty years I lived a lifetime working in or about the Climax Mine. For another ten, I watched it carefully from an office in Golden. Fifty years after I had gone to work and seventeen after I had retired, I wrote my first book entitled, *The Climax Mine, An Old Man Remembers the Way it Was*.

It was a collection of tales about the people of Climax and how they worked and lived. When selling the book, I became reacquainted with many of them and they told their personal stories to me or reminded me of characters or incidents worth a story.

The very first time I had a book signing, two elderly ladies come by and introduced themselves. "I'm Liz Schneider and this is my sister Cora May -- used to be Lee. We lived at Climax way before your time. Dad worked up there in the thirties. We lived in lower camp in a log cabin. It was pretty nice. I remember how mad I was when Bill Coulter tried to get some money from New York to build houses for the employees and he told them we lived in old shacks down by the tracks. I thought our cabin was pretty darn nice."

She continued, "We were waiting tables at the old White Level boarding house when they first started hauling ore out of the Phillipson tunnel. Our boy friends were motormen on the trains and when they were on graveyard shift, we would ride into the mine with them."

"Bill Coulter heard a rumor about it and came up to the portal one night and caught us. He cussed and raved and threatened to throw us in jail. Then he took us down to Daddy and threatened to fire him if he couldn't make us behave."

I could well imagine what a furor two girls riding into the mine during the thirties might have caused, on any shift, with anybody.

She looked at me with a twinkle in her eye. "We really had to be careful after that," she said.

One of the stories I had written told how George Schupkagel taught me how to tie flies for fishing. I received a letter from his sister, whom I had never met, and I quote in part.

"Hello Mr. Ludwig: We have just finished reading your book. Brought back a lot of memories. George Schupkagel was a brother. His son Butch's wife got the book for him. I have a brother Bill, our sister Jessie Francisco who is 91 lives in Leadville."

"In the 30's times were tough. George with a friend went to Colorado hoping to find work. In time my two brothers, my husband Jess and I went out. George was working in the mine at Climax. Jess got a job in the mine." ------- "When Jess's partner shooting down a hangup was crushed under rock, he quit. Then we came back home." George stayed with it, married, and had Butch and two daughters."

Everyone remembered a story.

Harry Smith said, "Remember that World series in 1957?"

Indeed I did. I was Muck foreman on the Phillipson and Harry was cleaning cars at the bins. I walked down the hanging wall, stuck my head in the Number One switch door and asked Beach if he had heard a score. He said that someone had called on the phone and Don Larsen was pitching for the Yankees, had a no hitter after five. He thought Harry might have a radio at the bins. I caught the next train out.

When we stopped at the portal I asked the motorman for his locomotive radio, pressed the button and announced, "Don Larsen has a no-hitter after five -- stay tuned." Then everyone in the mine anywhere near a loci radio knew the score.

Car Cleaner

Crusher

Harry did indeed have a radio of sorts. It scratched and crackled, but with careful attention I could follow the only perfect game in World Series history and report it nearly play by play throughout the mine.

Here is a short excerpt from the book of my arriving in Climax in June of 1950.

That evening as I sat reading in the Library, a rustic looking fellow sidled up to my chair. "Names Tennessee Tom. You ever play baseball?"

"Yeah, a little. Didn't bring my glove though."

"SsOK. Use one of mine until you see if you can make the team. You got a name?"

"Yuh, Jim, Jim Ludwig from Wisconsin."

"See ya around. We'll throw a little up in the Gym tomorrow night. Gorsuch promised me they would get a bulldozer up and clean the snow off the field this week."

I played every Sunday in summer for four years.

Tennessee Tom (Elmo Thompson) called me from Lakewood after he read my book. We talked about the old days and he asked if I might write a bit about the baseball team. Here it is.

Baseball at 12,000 ft.

Well, maybe 11,600. I know it was a little higher than the Phillipson portal. I'm not sure when the field was built, probably when Climax decided to make their town a desirable place to live. In 1936 Bill Coulter, in his recommendation for increased housing and recreation facilities in order to get enough employees to raise daily tonnage from 8000 tons per day to 12000 tons per day, mentioned "baseball grounds" as a necessity.

They found a spot east of the camp toward the Glory Hole where Ten Mile creek had become a dry streambed, and began bulldozing the dirt until it was too far to push. Some miners cribbed the uphill escarpment. Two by tens across the exposed crib timber ends made convenient bleacher seats. Some used pipe from the mine and some rock netting made a backstop. Cars could park on the hill above the field and along the road that passed along the right field line.

I'm not sure where they got the infield dirt, but it wasn't bad, once you got used to the ball skidding a bit. There was no grass anywhere. A high pop foul would hit up on the hill behind the cars and roll back toward the infield. The left field foul line was only about 250 feet to the fence, but the fence fell away rapidly until right center was quite deep. The fence was about four foot woven wire and if you went over it in center field you might wind up down a forty-foot rock embankment.

All in all, the quality of the field matched the quality of the players quite well.

We did have two things most parks of that day didn't have. There was an announcer's booth with a real loudspeaker right behind home plate on the hill, and lights -- lights enough to play softball if you were careful.

The summer softball games always drew a lot of interest. Different departments fielded teams to form a night league. The currently popular slow pitch was not yet played at that time. One night Flash Gordon, who pitched for the Mine, substituted a cantaloupe for the softball when pitching to Chet Burns who was Chief Electrician. Chet was along in years and had never seen such a soft pitch come from Gordon. His mighty swing connected perfectly.

The weather was sometimes a problem and night softball could be downright chilly. I remember delaying a Sunday afternoon baseball game in July until a snow squall passed.

Rotating shift work often interfered with your lineup. I was on a six-day schedule and was on Saturday night swing. About eight a.m. Sunday, Tennessee Tom called and said I would have to pitch because he had no one else who could throw hard. I was normally an infielder and hadn't pitched in years. Worse than that, my wife and I had gone to Leadville after swing shift and had unsuccessfully tried to close the saloons, which had swinging doors twenty-four hours a day.

Tom was the manager and catcher. He grabbed a mitt and said, "Let's see what you got." I said, "My curve won't catch and my fastball won't hop. That leaves me a medium hard one inside and high, and a damn slow one outside and low. I'll just try to find the plate with the rest."

We won that day.

In one game I hit two fly balls over that short left field fence and each was good for a free quart of Ray Schuette's home brew. We opened it right there on the bench that served as a dugout. When it about blew my head off, I knew why he was giving it away.

Our league was sort of loosely organized with Climax, Leadville Merchants, Leadville Giants, Salida, the Reformatory at Buena Vista, Dillon, Red Cliff, Minturn, Aspen and either Eagle or Gypsum. Some years most of them could field a team.

Red Cliff played at a flat spot along side of the Homestake Creek road. Cars formed the right and left field foul lines. Alex Ruibal, our center fielder, lost his temper at somebody ragging him on the sidelines. He started in from center field, intent on cleaning house, but I stopped him at shortstop when I saw his antagonist get a tire iron out of his car trunk. We weren't good, but we played with passion.

The old Aspen field was at the open spot near the bottom of their first ski lift. The old original Red Onion Saloon was in deep center field and there was no outfield fence. The bartender was sitting on the front step on a slow Sunday afternoon, when this guy hits a long one right past Alex that rolled right up to him. He picked it up and ran inside, Alex screaming right behind him. About the time the hitter was crossing home plate, Alex came out with he ball in one hand and an open beer in the other. Next batter up!

Dillon's field was along the old Highway 91 where it headed toward Kremmling. The picket pin gophers decided to build a colony on it and they would have to scrape off the mounds before a game. Ray Schuette hit a little Texas leaguer over second base. The ball bounced twice and disappeared down a gopher hole. First they couldn't find it and then they couldn't get it out of the hole. We all were screaming for Ray to keep running and he managed to cross the plate for an inside-the-park home run before the ball got there.

On the way home we stopped for dinner with my contract partner, W. O. Jones and his wife. Of course we had to stay and dance for a while. W. O. got in an argument about the game and invited his opponent outside to settle it properly with their fists. When he did come back in, he was a mess. I asked him what happened.

He said, "That SOB hit me only once. When I woke up all I could see was tires. He had dragged me over and stuck my head in between the back duals of an eighteen wheeler."

Carl Douglas was Warden at the Reformatory. (It really was one then.) He allowed inmates on good behavior to be on a traveling baseball team. They always had some very talented guys; the problem was the good behavior part.

I came up to bat one time and said to the catcher, "What are you doing here? You told me you were getting out last fall."

He replied, "I did. But I can always manage to steal just enough to get back in for the baseball season."

Their field was inside the west fence; with the north wing the left field wall and the west wing the right field wall. There were a lot of strikeouts because everyone was trying to break a window.

Actually, we had a game nearly every week and it was a great pastime. I think the league collapsed when Climax began to disassemble the town, but mostly because Sunday small town baseball just went out of style. I never played after 1953. We decided fishing and the outdoors was better family recreation.

jimmyj 3/3/02

Someone said, "You have a bit about Idas Price. Why don't you tell the story of why he came to Climax?"

Here it is.

Idas

We crept silently into the open cave above 240 South. Idas wanted to show me a stub pillar that was crushing the concrete above a loading cutout. The only light came from our cap lamps. I held mine in hand the way muck hands do, but Idas left his on a full brimmed hard hat, the way miners do, and slowly swiveled his head in the direction he wished to see. We did not speak aloud, but wiggled the spots of our lights to indicate a point of interest.

"Thirty cases right there," he whispered, pointing his light at a spot where the arching back rested on an apex over the haulage. The rock was obviously crushing at that point.

I moved my spot back and forth to indicate I did not approve. His light traced a hump-backed arc through the blackness, signaling, "Let's get out of here! "

We carefully threaded our way along the apex to the safety of the open undercut between two unshot pillars.

"Why not, Jim? We may crush that cutout if we don't blast that stub."

"I won't let you take any men into that cave, Idas, lose the cutout or not." I said. "We might get somebody killed."

We were deep in thought as we walked to the lunchroom. Talking mostly to myself, I asked, "Now why in the world didn't that pillar blast cleanly when we shot the longholes? I know it was drilled and loaded well."

"Let me show you," said Idas. He set his pie can on the floor and kicked it over. "See how easy that was?"

He set his pie can upright and put his foot heavily on top. " Now try to kick it over," he said.

I could not.

"We have to find a way to blast the pillars before they take weight. It is as simple as that."

Hog Farmer Engineering, I called it, and how often it has taken me through a hard spot. But Danny, Slim, George and I packed two very lifeless bodies from a similar cave before we developed a method of fan drilling from a subdrift to do just what the hog farmer had suggested. The method was refined to mine the rest of Climax, the Urad and Henderson mines.

Idas must have blown into Climax about 1952. I know they were still contract mining when I first heard that squeaky voice rise higher and higher as an argument progressed. The argument might have been about anything under the sun, or in the mine for that matter, because Idas had an opinion on everything. He usually had the money to back up his opinions and would bet on anything.

He had been a poor Oklahoma farmer, among other things and someone who had to scratch for a living at one time often made a very good miner. And he was a good one; there was no doubt about that.

I really didn't know him well until we were battling the heavy ground in the 240 area. I had returned to the Phillipson level after a lengthy stay on the Storke, then a while on Phillipson production before becoming a General

Foreman. Idas had become a shift boss and he made it his avocation to teach me how to mine stopes. The fact that I had been a stope miner before he ever saw a rock drill didn't make any difference.

When I moved to Buena Vista, we traveled together in a car pool and became quite good friends -- at least as good as you cared to be with a man of his reputation. He loved to honky-tonk, knew every barfly, every barmaid and every shady character of the underbelly of the culture of the western mining towns we inhabited. He was everybody's friend and would give you the shirt off his back. The women loved him and a long-widowed bartender once told me Idas was the best in town, what ever that meant.

One day in the carpool I asked him why in the world he ever came to Climax. He mumbled something about states evidence in a fraud trial and thought he better leave Oklahoma, then changed the subject.

That really piqued my curiosity, so when we stopped by the Green Parrot for a beer, I asked him again. He motioned me over to one of the booths in front of the bar, bought another beer and we sat down.

"It's a long story," he said. "You would never know it, but I used to be a God-fearing, church going, Baptist family man. We married young, and were dirt-poor farmers. It was a hard life. The depression was barely over and I finally got a part time job selling candy and nuts to country stores and taverns for a well-known candy company. It didn't pay much, but it helped a little.

"My boss had been around, even to Saint Louis and New Orleans, which made quite an impression on a country boy. He introduced me to beer, hard liquor and wild women. I discovered I loved all three. Especially the women. I learned that a country barmaid was just like a candy bar, sweet and delicious once you took the wrapper off.

"I began to ignore my wife and farm, but she took care of things. I don't know why she put up with me as long as she did, but she finally had enough when I was shot."

"How did that happen?" I asked.

"One day after I had coaxed this gal into bed, her husband arrived home. I grabbed my clothes and ran out the back door, headed for a horse barn thirty yards away. The first shot slammed into the side of the barn but the second hit me in the back. I staggered through the door and fell helplessly into a horse stall.

"When he found me, he put the gun barrel against my ear and pulled the trigger. The hammer clicked on an empty shell casing. He swore, spun the cylinder, swore again and walked away."

"I'm not sure how long I lay there before the sheriff picked me up and took me to the hospital. The slug had lodged in my liver and they said it would be more dangerous to remove it than leave it there. That's why I flunked the Army physical when the war started."

He motioned to Lloyd Hooper who had just walked in the door. Hoop slid into the booth next to him and Idas said "Hoop why don't you help me tell Jim about the soybeans? You were there and remember better than I do."

Over the next hour and a few more beers Idas told me this story.

"Times were hard for us Okies just before the war. The drought wasn't really over, crops were poor, and many of our kin were Californy bound. That wasn't any better, so I heard. Selling candy helped a bit, but then the war came on and Uncle Sam kept all the Baby Ruths and O Henrys for the troops and we could hardly sell the other junk we had. Oh, they would slip us just a few, we could have jacked the price sky high, but you remember how the OPA, the Price Administration boys, kept an eagle eye for wartime profiteers.

"Even peanuts were impossible to get, but then they started roasting soybeans, bagging them in pretty cellophane bags. When they were fresh they tasted good and if you didn't crack your teeth you could get by. But setting on the back bar in the Oklahoma heat soon made them rancid and they spoiled.

"Our customers asked us if we'd take them back and we said yes, but only if the company would refund our cost. We sent some to Tulsa and the company said, 'Keep the stinking things, just let us know how many you take back and here's a check for what you got so far.' We just stacked them in the shed after that.

"The boss wrinkled up his forehead, sent us traveling with a real deal: buy three boxes of these tasty beans, we'll sell you half a box of Baby Ruths, throw in a pack of peanuts now and then. You'll have to give us back the soybeans if they spoil.

"Not a bad deal for a country boy! We sold the beans and Baby Ruths; they gave us back the beans because they wouldn't sell. We got a refund from the company.

"I snipped the plastic bags in half, emptied out the beans and fed the pigs, who didn't care if beans were spoiled or not. I noticed most of them just passed right through.

"One day a feller says, 'Why go through all the trouble delivering all those beans? Just put them on the invoice and keep them in the truck. No one will ever know.'

"I wound up with cases in the shed, so we bought an old hammer mill to grind them up and did the pigs ever get fat on that soybean meal."

Hooper chimed in, "You wouldn't believe it, Jim. When they were milling soybeans, the cellophane wrappers would blow in the wind and pile up against a fence row just like tumbleweeds."

Idas continued, "Then we bought some more pigs and really began pushing our roasted soybeans. We would order in 50 cases and take them directly to the hammer mill to feed the pigs. Then we would invoice a business for them if he wanted any candy, but he never saw them. Then we would get a refund from the company for spoiled beans. We bought them once and sold them three times, including once as pork.

"We might have kept going longer if the boss hadn't bought that damn pink Cadillac to sport his girlfriends around. I took care of the pigs, but he made me give him half the profits. I'm not sure who turned us in to the Internal Revenue Service. I'm sure the company must have been suspicious when we bought more soybeans than the rest of the dealers put together, but I think it was probably a jealous girlfriend.

"Next thing I knew, we were in jail, charged with defrauding the company and not reporting the income for taxes. Hell, I was so scared that when they offered me a plea bargain, I squealed like a stuck hog. Wound up with probation, but my boss spent time up at Leavenworth in the federal pen.

"I had met and married Sally by that time, and as soon as I got off probation, I came to work at Climax. I knew it was time to settle down anyway."

"Yeah, sure," said Hooper.

I believe Idas left Climax sometime during the '70s. I know he wound up at the Missouri lead mining operation where he worked for Charley Johnson, who had followed Ralph Barnett, Don Glover and Ed Allen to Missouri. I received a letter from Charley recently, noting that he had attended Idas' funeral.

You had to know him to believe him. He was truly one of the characters that made up our life and times at Climax.

jimmyj 1/12/05

Frank Cerise, my father-in-law, was also one of old time characters and I had written about his trip to poach some venison on the Frying Pan River. By the time I knew him he had gained a reputation as a spy for the Fish and Game, some said actually a warden. One year he had them put some fish in the little pond over on Mount Democrat where he dutifully fed them for a year or so. Just when they were the right size, some miner dropped a stick of dynamite in the pond and killed them all.

Here is another story about him.

Grandpa and the Skunk

The Climax exploration guys were drilling in the Twin Peaks area back in 1958. They found molybdenum, didn't look too bad, in fact it looked as if they might have something there.

Little Bob decided that before the word got out they ought buy some water in the area, just in case the prospect was for real. Jack Laing, the Leadville lawyer, never one to miss a buck, offered Bob his cabin, with his land and water at Twin Lakes.

The deal was done; Bob had to find a caretaker to watch the place. Why not his old friend Frank Cerise? Frank could garden with the best and was cantankerous enough to take good care of things. So in the year of '59, Grandpa Frank and Grandma Frances moved down from Climax, where they'd lived for nearly twenty years.

Soon the place was lawns and gardens, flowers blooming everywhere. Grandpa worked from dawn till dark; he never could sit still. Grandma pitched in too, they really loved it there.

When the winter came he shoveled snow, fed the birds, and even cleaned a pond for skating for the kids.

One winter day as he complained there just was nothing else to do, I suggested that he learn to trap the bobcats that seemed to wander all about, catching snowshoe rabbits in the swamp. The price for fur was very good. When I remarked that over forty inches, nose to tail, was a pretty good-sized cat, he had it cured and it wound up on my wall. It still hangs there. He trapped for six-eight years, until the protesters knocked down the price of fur.

One winter he had set a trap next to a pond, down at Dr. Lipscomb's place. He placed it where it could be seen from on the road, as he drove by to get the mail.

The snow squeaked 'neath his boots, this winter day, as he went out to start the car. The smell of skunk was strong; it wafted up from down below. He guessed that one had strayed into his bobcat trap. Sure enough, he could see it moving at the trap. He didn't have his gun, so he decided just to get the mail and then come back again to shoot the poor damn thing and bury it somewhere. It was another half mile into town; the gossip of Twin Lakes took just a little time.

Now Grandpa had a neighbor who lived about a mile further up the road. She complained of Grandpa's traps, made him move one once, when he made a trail across a corner of her land. Grandpa told her where to go. "There wasn't any trapping there," he said. "The fur was bad because climate was too hot."

This day by happenstance, she followed Grandpa into town. As she drove by the Lipscomb driveway, a movement caught her eye. She stopped the car, walked over to where she could clearly see the skunk caught in the trap. She just had to let the poor thing go.

No one knows what happened next, but they smelled it up and down the valley. When Grandpa drove back by, he saw the fracas. He didn't stop, but drove on home. The Warden lived in Leadville, so Grandpa called him, explaining how someone was tearing up his trap line, and if the Warden came right down he thought he had the goods on her. And yes, he would file charges if he could.

The Warden met with Grandpa at Dr. Lipscomb's gate. They saw the tracks of someone, saw the skunk was gone, smelled the stench, the snow looked like a wrestling match had just occurred.

It was obvious that Grandpa had a case.

They drove up the road and stopped beside the neighbor's car. The smell of skunk was everywhere. When she answered to their knock, the smell within the house was just as bad.

The warden read the law aloud, and then he said to Grandpa, "Do you want to file charges against her?"

"Yes, I do," said Grandpa. "I'm tired of her messing with my traps."

The warden was perplexed. "Maybe if she paid you for the skunk you lost, we could forget about it and not cause a great big fuss. How much do you think that skunk was worth?"

With a face so straight and serious, Grandpa Frank replied, "I could have gotten fifty dollars, easily, with the price of fur the way it is."

She scrambled for her purse, and held fifty dollars out to him.

"How about it, Frank?" the warden asked.

"Oh, I suppose," said Grandpa. "But, by God, if you ever touch my traps again, I'll have you thrown in jail until you rot!" He took the fifty dollars, and placed it in his wallet as they went out the door.

When they drove up the road, the warden turned to Grandpa, "Frank, you are an ornery old cuss. You didn't even thank her for the favor that she did. You won't have to bury that damn stinking skunk yourself."

jimmyj, 10/18/97

The Town of Climax was the entity that seemed to provide the soul of what was really just an ordinary company. Where management had been encouraged to live at the town site until about 1956, Management was encouraged to lead the move away from Climax when it became obvious the business would not support the paternalistic policies that had been the norm since the thirties.

I recalled the old miner's barroom lament. "Someday I'm gonna leave This Place. I'm gonna go to where the grass is always green and the snow is never ten foot deep. Yes sir, someday I'm gonna leave This Place. But first, let's all have another beer."

Each of us who lived there and worked there had each expressed that thought at sometime or another. We never dreamed that someday This Place would leave us.

Now it is gone. There will never be another.

Chapter 6, Evergreen and Buena Vista

Let me return to the beginning of this story when I was asked to move beyond the Glory Hole, to Western Operations headquarters in Golden. In many ways the years I and my family lived in Buena Vista, between 1958 and 1972, while I was in Climax Mine management, were the best years of our lives. The decision to move was a difficult one. Twenty years ago I had dreamed "running this place" as the Resident Manager. Now I did.

Yet, the opportunities in Golden seemed limitless.

It had become obvious that the Climax mine would not support the paternalistic policies that had been the norm since the thirties. Now the town was gone and an open pit would soon swallow the Glory Hole. The Glory Hole and the Town of Climax were the soul of what was really just an ordinary corporation.

My wife didn't want to leave Buena Vista, and refused to move to the city. Evergreen appeared to be a workable compromise and we moved there in midsummer 1972.

We attempted a new lifestyle, and for the next ten years we struggled to adapt, but never succeeded. When we returned ten years later, it was as if it had been an overnight stay at a hotel. Buena Vista was home, and always will be.

Some of the challenges of my new job were unanticipated. When I learned my family was expected to entertain company officials and guests at home, I simply refused and stuck to it. When I was supposed to get an MBA from Harvard, I refused to be away from my family for nine months. Gouseland, who was then president of Climax, said, "Everyone says his or her family comes first, but you really mean it, don't you?"

"Yes," I said. "I have three teenage boys and if they don't need me now, they never will."

When I was asked to move to New York or Canada, I said no. When asked to say something nice about New York, I said the best thing about it was that it was a long way from Denver. I fought headquarters and tried to keep them out of our operator's hair. When I was asked where Amax headquarters was located, I said, "Just follow the Yellow Brick Road."

In Golden the opportunities to enjoy life and not take yourself too seriously were considerably less than when working at Climax, or as we commonly said, "On the Hill."

I was able to convince Terry Fitzsimmons to move to the Public Relations Group in our office, and he also located in Evergreen. Our two-man carpool to Golden provided a sounding board for each other and helped us to keep our sanity.

We began a new life. Many weekends we would go to Buena Vista and work on the nursery -- the "farm" we called it. While I was building the first fences an amusing incident occurred.

Kevin and the Bear

This story unfolded in the spring of 1972 when Buena Vista was a sleepy little town.

The bear had worked his way up from the south, from Chalk Creek north to Cogan's, and stayed in the piñon pines west of the river where the airport is now located. Each time he wandered east, there was a highway where the cars buzzed back and forth. He continued north to where a bunch of cars had stopped along the dirt road south of town to see that rarity, a real bear.

The bear had never seen such things, had no idea what the cars and people were. The smell of cows was more familiar; he knew they wouldn't bother him at all.

He drifted to the west, away from all the people on the road. He crossed the irrigation ditch, crawled through a fence into our pasture, where I was keeping about forty Holstien heifers.

Gary, Steve and Mark had helped Jim Mahon and I build a garage to store things we did not wish to move to Evergreen. For their help, I had promised to buy them a dirt bike. Jimmy had chosen the new yellow Kawasaki bike, and we had marked a track to the south, on the dry land meadow that we called the mesa.

The mesa was a flat with cactus, rocks and grama grass, fragile vegetation at the best, and we needed to limit damage from the tires.

The bike track was a triangle; the north leg paralleled the Cottonwood irrigation ditch, which rushed its precious cargo of water fifteen feet below. Kevin Farrar, a neighbor Gary's age, also had a motorcycle that he was allowed to ride on the track.

That night, as we left work to go to my daughter Linda's, he remained, roaring round and round the track. I saw a police car rushing by and joined the other cars that pulled into my farm driveway to see a rarity, a bear.

The bear ambled through the pasture, loping right along. The cows just sort of stepped aside to let him pass. He glanced our way, and then angled toward the ditch and mesa to the south. The bear could not see over the rise beyond the irrigation ditch as he splashed through it and up the hill.

Just then Kevin came around the curve and opened up the gas. He and the bear met face to face.

Kevin jumped free of the bike, fell and scrambled to his feet, the bike still snorting, spinning round while lying on the ground. Kevin ran for home, a quarter mile away. He never did look back.

The bear was not about to face that horrible machine; he ran back down the hill, through the irrigation ditch, crossed the meadow just beyond we spectators and ran into town.

It was a funny sight. I don't know who was the most frightened, Kevin or the bear.

We have laughed about the incident every time it is recalled.

The times have changed, in a direction we would not have guessed. In 1972 a crowd would gather at the sighting of a bear, and deer or elk were seldom seen near town.

Now we must fence the nursery eight feet high against the deer, the elk herd of 800 winters north of here and bears are not uncommon in our neighborhood. Foxes raise their kits beneath the old abandoned Turner house, and clean my gopher traps each night in fall. Deer graze in our flowerbeds and browse on lilac bushes in the yard.

We must be careful to keep garbage under lock and key and be careful when we're hiking because mountain lions have followed the deer into town. I can often show you mountain sheep and sometimes mountain goats not far away, and have pictures of antelope within a hundred yards of my front door.

Our little country town has grown haphazardly and homes are common through the hills and flats. The animals have moved right in to live with us. A modern bear would pause and let the motorcycle pass, then continue on his way.

They say that we have ruined their habitat. That might be, but they have taken over ours.

jimmyj 2/10/99

My nursery at Buena Vista was a welcome respite from the growing pressures of Western Operations. I was learning to propagate native, high altitude plants and found that I was capable of landscape design. Income from mine reclamation was sporadic at best. My youngest son, Gary, began to manage the operation in 1978.

Here are a couple of stories about that time.

The Four-Mile Ranch (first published by Colorado Central Magazine)

One day I was scrounging through a box of odds and ends, Dad's jewels, the kids called them, and I found an old chert Indian hide scraper. It fit comfortably in the palm of the hand, the scraping edge sharpened by careful chipping of the chert. I recognized it immediately. We had found it in the early sixties on the sunny side of a rock cliff on the Four-Mile Ranch. The kids and I had gone there to spend a bright, warm March day looking for Indian artifacts, and this was probably our best find.

It brought back a flood of memories of that place and those times. Let me tell you a tale.

I believe the year was 1959; I know it was a long time ago. We lived in the little log house on James Street. Gary was a baby, and with Steve and Mark, Deb and Linda, Mom and I, we filled the place.

Next door lived Mrs. Cyr. That lovely lady watched us carefully, to be sure that we would get along and feel welcome in her town.

The house north of us was empty, an old house. I looked in through an open door. The plaster had pulled from the lath in spots and lay scattered on the floor. A white clapboard house, known as "The Brown House," it looked barely livable.

One day a family moved in. One look at the man and I knew he was no farmer just blown in from Oklahoma, to give the mine at Climax a try.

A sweat-stained Stetson sat squarely on his head, the right hand roll dirty with sweat and grease, the kind of dirt that comes from jamming it on when working on a baler or branding a calf.

He wore a long sleeved shirt -- three snaps held it snugly at each wrist -- and tightly fitted Levis with a wide leather belt. His boots rolled slightly out and were worse for wear.

The wind had burned the wrinkles on his face; the beard was grizzled gray, although he wasn't an old man. Had I never known a real cowboy, I would have guessed, this must be one.

We waved a hand of recognition, walked toward each other, trailing kids behind. We met at the fence, in perfect timing each reached out a hand.

"Ray Sailor." he said firmly. He had a solid grip, from a worn and callused hand.

"Jim Ludwig," I replied. "Welcome to the neighborhood." Steve tugged at my leg. "And this is Steve, and Mark and Linda. Say hello to Mr. Sailor, please." Very solemnly, they did.

"This big girl here is Joyce," he said, "and Ruth, and Randy. Say hi kids. Looks like Randy must be Linda's age, How about it, Son?"

"I'm six and a half," he said.

"Me too," said Linda, quite defiantly.

They were a pair to draw to, that was evident immediately.

"I'm a miner up at Climax," I explained. "Moved in last summer after eight up on the hill."

"We just bought the Chicago Ranch," he said. "Sold my place in Kansas awhile back. It was too small to make a living, but too big to pay the taxes every year."

"You've a lot of guts," I said. "This ranching business is tough. Let me wish you luck. How come you moved in here?"

"The Chicago Ranch house is a shambles," he replied. We need to build it over. It needs a bathroom and a kitchen and lots more. Bring the Missus over, meet Inez, my wife. Since we're neighbors, we might as well be friends."

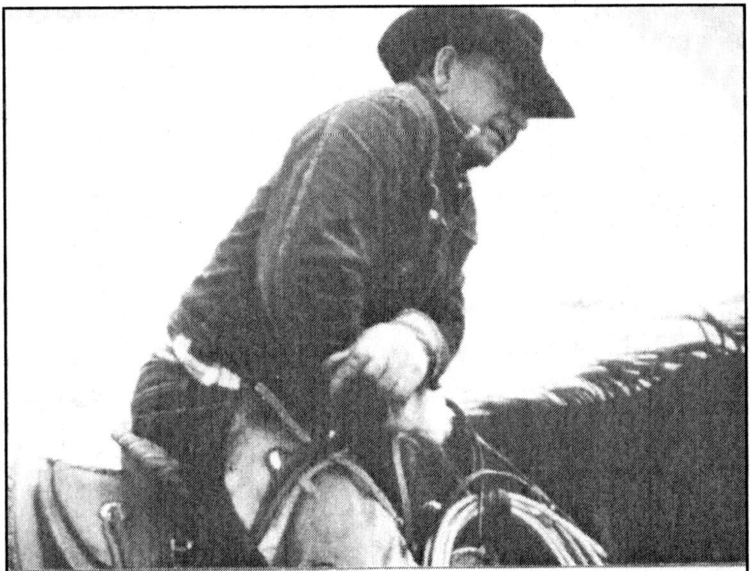

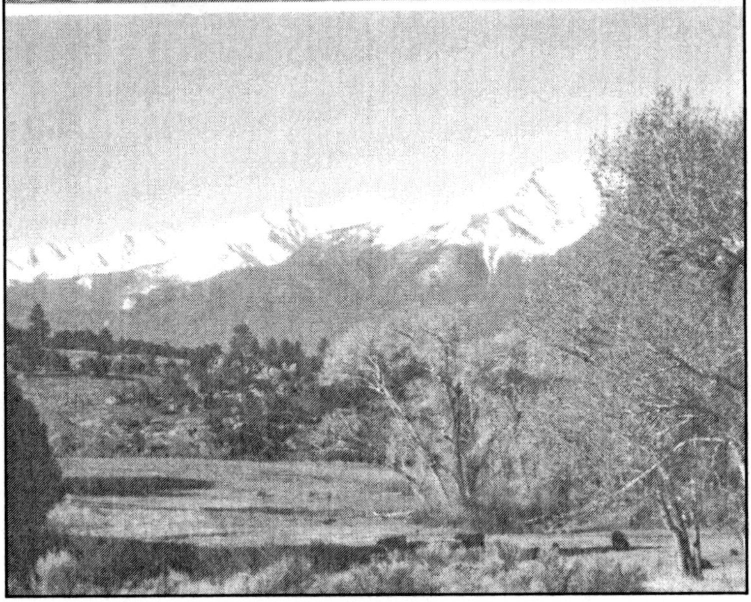

And we were. Immediately. There was something about Ray, his grip and friendly, honest eyes that made you want to have him on your side.

We were busy raising families, with long day jobs, but we got acquainted anyway. One Friday night we had a beer, and Ray said, "Why don't you ride up to the Four-Mile Ranch with me? I have to irrigate. I'll head out about daybreak. You're going to like that place."

I had only seen it from the piñons up above. I couldn't wait.

The light of dawn was barely peaking through the ponderosas on the Jordan place across the road when I heard the pickup's horn. The truck had seen its best miles years ago, banged and rattled as traveled on the washboard gravel road.

Ray told me how the Four-Mile, and Pyle's pasture just across the hump, controlled the grazing rights on the east side of the Arkansas river. He told me how the ranchers would protect a herd from winter winds on the ranch, and how the cows would calve in comfort and seclusion.

We crossed the Arkansas and as the old railroad tunnels came into view, we took a hard right up the hill. We bounced around the rocks and piñons, past the fire pit where now and then the high school kids would tap a keg.

"This trail to the right," he said, "goes down to the arrastra by the creek, where old-time miners ground the ore to look for gold. If you wander far enough, you'll find the Goddard homestead and the Seven-Mile Pass. The Indians used it to come into this valley from South Park, rather than the Trout Creek pass we use today.

"Since ancient times the Indians met to trade in this part of the upper Arkansas that we call the banana belt. Arapahos, Comanches came from the eastern plains, Pueblos and Navajos came through the San Luis Valley into this, the homeland of the Utes. Many must have camped on Four-Mile. There are arrowheads and flint chips scattered about. The rock they used to make them isn't native to this valley, and has been carried in from God knows where.

"This next road by the cattle guard would come out by the old corral, but it is washed out. We'll go on up above and drop down from the top. The irrigation ditch out of the Four-Mile Creek is up above the meadow, anyway."

There was a switch back and a curve or two, and then we cleared the piñons for a bit. The valley of the Four-Mile was on our right, still partly hidden by the gloom of the not-fully-risen sun.

The ancient rocks beyond, in sculptured beauty, cast their lengthy shadows on the valley floor. To the north, the twin Buffalo peaks glowed brightly in the sun, the white of winter snow now nearly gone.

We turned to the right, the road reduced to double tire tracks, and drove down to the padlocked gate. As I recall it wasn't long before Ray threw that lock away.

The meadow lay before us; a thin line of willows followed the life-giving stream. Some Hereford lowed across the way, content to feel the warmth of the rising sun. The green of irrigated grass was bright, but above the ditch, the gray-brown of the desert stretched endlessly.

We trundled across the creek, through a grove of cottonwoods and parked above the ditch, where a pair of canvas dams kicked out the water, the life giving blood of the land.

As Ray rearranged the dams, I stared in absolute delight at the beauty of the place. My back was to the rising sun; before me stretched the valley of a drugstore cowboy's dreams. To the right, about one half mile, the meadow pinched back to the stream. I could barely see a cabin to the left before the rocks cut off the view.

But lifting up my eyes above those rocks, and looking far across the Arkansas valley I could see the peaks along the continental divide, white and bright with snow, in shining glory, in deep contrast to the blue of mountain sky.

I recalled another traveler, who years before, had looked down upon another valley, in another state, and exclaimed, "This is the place!"

But to call this place a ranch was stretching the term quite a bit.

The little cabin smelled of pack rats. A porcupine in endless search for salt had chewed a hole through the rough-sawn floor. Only a windowpane or two remained.

"Someday I'm going to fix it up," Ray said. "We can camp out here when we push the herd up the valley, through the pass into the meadows of Rough and Tumbling Creek and to the grass that grows there. I have grazing rights for eighty pairs."

A lean-to had been built against the rocks, near a barn of large hand-hewn logs. The chinking had weathered away; the slabwood was gone to decorate a city den. The old corral had been repaired enough to hold some stock. The creek was nearly dry, its water stolen to green the meadow above.

Not a ranch really, just a place to dream.

We worked our way back through the meadow. Ray with practiced eye, checked the cattle carefully.

"You're welcome anytime," said Ray. "It is such a peaceful place. Soon we'll move the cattle up the hill, then you can have the run of things."

We took him at his word, and for the next few years we camped or picnicked, and searched the hills for arrowheads.

I watched the cabin be repaired. The fences, the corral were put into use again. The washed-out road was soon repaired.

The Sailors moved to the remodeled house at the Chicago Ranch on Rodeo Road. Inez told me they had to move out there. Randy had to have a place to run. He felt caged in town.

We moved across the street into the Jordan House, and raised our family there. The kids grew up, and we sort of grew apart.

Now and then Ray and I would have a beer at the Hotel Bar. We'd catch up on our families, then straighten out the problems of the world a bit. We had a close and casual friendship; we didn't need to talk to keep it lit.

He was a friend like no other.

One Saturday, not long before we moved away, I drove up to the Four-Mile and found a big new lock and a no trespassing sign. Now what the hell?

I came back to Main Street, saw Ray's pickup at the old Hotel.

I slowly entered and looked down through the crowd and hazy smoke; saw him sitting at the bar. All alone, he was staring at an empty shot glass, empty beer glass, another half-filled beer.

I motioned to the barkeep; she filled the shot glass, set me up a beer. I straddled the stool next to him, "Hello Ray. Long time, no see."

"Hello Jimmy," he replied.

He raised the shot glass to my beer, "In your eye!" he said, then tossed the whiskey down.

"You sell the Four-Mile Ranch?' I asked.

"Yep."

I took a slow drink of my beer. "Who to?"

"Some guy from Chicago." His eyes were squinted, almost closed, still looking at the beer glass in his hand.

I raised a thumb to the bar tender. "Better fill them up, again."

I looked back at Ray, "Now why the hell did you do that?" I asked. "He'll probably fill it full of summer homes for city slickers from back east."

"I know, I know." His voice broke just a bit. "I had to, Jimmy. It was either sell that place or lose the whole Chicago Ranch. It was the only way that I could raise the money that I had to have." He downed the shot.

I caught the bar girl's eye, "Better give me a shot of that stuff," I said as she filled them up again. I threw a twenty on the bar. "Thanks."

There was nothing else to say. We sat silently, staring at our beers. "This one's on me," she said.

I slid off the stool and extended my hand. His hard and callused hand squeezed my office-softened palm.

"Oh, crap!" I said, choked and turned and left the bar, left him staring at his beer. I did not wipe the tear until I was out of sight.

That was the last time that I saw my friend.

Epilogue

Some years later in my office at Amax headquarters in Golden, my chief engineer, Max, and I were meeting with some land men from the U. S. Forest Service. We had a large-scale map of the State of Colorado spread across the desk. The National Forest lands were in bright green, the Bureau of Land Management land in yellow and the privately held land in tan. There is a lot of public land in this state.

We had just agreed to see if we could buy some isolated homesteads near the Cochetopa Pass in the Rio Grande National Forest northwest of Saguache. They were near the old Ute Indian agency and were owned by the Coleman ranches. Like ranchers everywhere in the late seventies, the Coleman's were having trouble making ends meet and the Forest Service wanted to return those lands to the forest by exchange for land we needed at our developing mines.

Our discussion of the plight of the ranchers reminded me of Ray Sailor. I pointed my pencil at a spot on the map just outside Buena Vista in the San Isabel National Forest, a tan thumb surrounded by green.

"Would you be interested in this?" I asked.

The Land Man squinted a bit, "Hell, yes," he said, "That's the old Four-Mile Ranch. About 1300 acres, used to be part of the Chicago Ranch. I heard some guy from back east is going to develop it. I don't believe it is for sale."

"We'll see," I replied.

After they left, I told Max, "You buy that land; whatever it takes."

He was able to do just that and the Four-Mile was exchanged back to the National Forest as a minor part of a larger exchange.

I retired to Buena Vista in the early eighties and I never did drive up to the Four-Mile because I wanted to remember it like it was twenty years before. Finally about 1986, I took my llama, Thistle, for a walk there. There was no trace of the barn, corral or cabin. All the fences and the ditches that defined the meadow were gone, the grove of cottonwood nearly dead from lack of water. Campers had built rock fire pits along the creek where it wasn't completely overgrown by willows. ATV trails went everywhere, except where the Forest Service had cabled off a portion of the upper meadow.

I liked it a lot better when water ran in the irrigation ditches, the grass was green and Herefords lazed on the hillsides.

Ray died suddenly in 1971, not long after I had talked to him in the Hotel Bar. I see Inez now and then; she lives by herself right near our driveway. She doesn't go to the Four Mile because the last time she did it made her cry.

Randy has developed a little place on what Ray called Pyle's Pasture over the hill east of Buena Vista that we call the Sleeping Indian. He runs a few cows and I believe still holds some of the grazing rights. Joyce has a home on the quarter section Ray was able to hold south of the Four-Mile near the Arkansas River, and tends what is left of the old Chicago Ranch.

And I sentimentally write of old times, old places and old friends. I shared this tale with Joyce, Ruth and Inez the other night. Joyce commented, "I always knew Dad felt very deeply about you, but never understood why. I didn't realize you felt the same way."

He was truly a friend like no other.
jimmyj 3/17/1999

Why we Bought the Llamas

It was in 1981. We were coming down the trail from Willis Lake and had stopped at the halfway meadow to catch a drink and take a break.

This was year 23 since I had started on the annual sojourn that had become tradition for my family. My three sons, Mark, Steve and Gary, nephews Jim and Donald and myself made up the crew.

Willis Lake

We had packed everything we needed on our backs for a two-night stay up at the lake. Fishing wasn't great. I have a picture of just nine, but two were eighteen inches more or less, the last rainbows out of Willis Lake. There would be only natives from then on.

The little spring that gave us sparkling water at this spot, flowed just as clearly as twenty years before, and maybe several hundred before that. The water tasted great, but I was hurting; every bone I had ached. Sitting in the office in Golden every day, out of shape and not so young, this trip became a chore for my poor body. The fun was gone, the pain had begun.

As we relaxed and drank our fill, I said, "This has been lot of fun over the years. But I tell you what, this is the end. I no longer can take fifty pounds upon my back and hike six miles up the hill. I remember when I carried all we'd need for Gary, Mark and Steve and me.

"Gary was only seven then. Each year you guys could carry a little more, and we've had some horses now and then. Remember a few years ago, Georgie Webster packed us in? Then that colt, his pride and joy, spooked and ran into the rocks and broke his leg. We packed all of our gear out on our backs again.

"I hate to say it, but I've had enough. If I don't quit, someday this hike will kill me. You'll pack me out of here, tied across a horse, or swinging from a chopper in a basket on a line."

"Aw, come on Dad," Mark said, "As soon as you get down and have a beer, a little rest and you will plan for next year once again."

"I've made up my mind, there'll be no more." I said, "Of course if you want to carry all my load, I just possibly could make it one more time."

We picked up our packs and headed down the hill.

About a month had passed when my daughter Kedran, who was living in Denver at that time, came to Evergreen to visit Mom and Dad.

"Dad," she said, "Mom and all us kids have an idea. When your birthday comes October 29, we want you to be a sport and play along with us to have some fun. Will you let us blindfold you and then go with us for a ride?"

I looked around the kitchen. Every eye looked right toward mine. "Now wait a minute; let me think. Surely you don't plan to put my feet in concrete and throw me in the drink?" I asked.

"Of course not, Jim," said Mom, "We're just trying to have a birthday party that you won't forget."

"Please, Dad," said Chrissy, "Just this once?"

"Oh I suppose," I said. "It does sort of sound like fun."

It was early on a Saturday near the end of October. Mom, along with my daughters Lissa, Chrissy, Debra and the driver, Kedran, waited at the station wagon with a blindfold just for me.

"You could give me a clue to where you're taking me," I said.

"OK," said Kedran, "We are headed down toward Denver. I won't tell you any more. So just relax and ride; we'll let you know when we get there."

Kedran cautioned everyone to watch their talk, to not give anything away. I thought that they couldn't lose me for too long. I knew that I could tell the freeway, or the downtown sounds.

But Kedran took me through some parking lots and alleys on the west side where she lived. Soon I was lost. It seemed like we were driving in the country for a long while.

At last we slowed down for a turn off of the blacktop, into a gravel drive. I heard the sound of animals that I had never heard, and smelled a distinct odor, that some would call a stench.

Kedran stopped the car, opened the door and let me out. She jerked the blindfold free.

"One of these is yours," she said, "We all agree. There is no way you can get out of this."

We were standing in a farmyard at Boulder. Forty llamas, more or less were wandering about. A pleasant looking, gray haired lady walked up and extended her hand.

"I'm Bobra Goldsmith. I am very pleased to finally meet you," she proclaimed. "Your wife and children have decided you must have some llamas on your farm. So I've gone to Buena Vista to take a look, and met with Gary there. It's a great place for llamas. I will be glad to help you all I can. So make yourself at home, and get acquainted with a very different kind of pet."

That we did. The llama craze had just begun and there weren't a lot of books to read. We even started a scrapbook to collect the things we learned.

Kedran's friend, Lee Ann, painted me a picture that still hangs upon our wall, and a stuffed toy about eighteen inches tall of a llama was mine to pet until the real ones came along.

Bobra helped a lot, we visited frequently and got a feel for what we could expect.

Finally, we decided to get three, a female to breed and two gelded males to use on packing trips. We first picked out a female named Frolic. She was white except for black ear tips and tail, with a little spot of brown atop her head.

One night we had a call from Bobra. "We've just had a baby llama and much to our surprise, he is a throwback to the wild quanacos of South America. We didn't realize his mother, whom we'd purchased from a Zoo, had quanaco blood at all. You had expressed an interest in that coloring before, you may have him if you wish. Would you like to pick his name?"

"I'm sure we would," I said, "My kids all are here. I'll call right back."

We were sitting at the kitchen table, drinking cider from a gallon jug. I told the kids what Bobra had told me. One of the little girls asked, "What color will he be?"

I pointed to her cider glass, "Just about like that," I said. So we named him Cider and I called Bobra.

Bobra picked the third one, a chocolate brown named Thistle, who was old enough to pack that year.

I built a pole shed on the farm and covered it with tin left over from the nursery building. The fences and the pastures all were done when Bobra brought the llamas to our place in June.

We worked with them all that summer, training them to pack. We assembled all our camping gear to fit the panniers on the llama saddles that we purchased.

And I, with nothing on my back, with Thistle packing all my gear, went to Willis Lake that year. I was able to continue making the trip for another ten years.

The tradition of spending the third weekend in August at Willis Lake continued through 1992. That year our crew consisted of three sons, three sons-in-law, two grandsons and myself. We loaded all our gear on the three llamas and stayed over three nights.

At the Clearing 1992

From 1957 until 1992 -- a span of thirty-five years, the lake, the trail, and our campsite had changed very little. There were many more fish in the lake, but they were smaller and all supposed to be the "native" native cutthroats now planted by the Fish and Game Department.

An incident occurred on that last trip which prompted my son-in-law, Jim Mahon, to say this was the only time in his life he wished he had a TV camera.

I was wakened in the night by the llamas' warning whinnies, so I crawled out of my sleeping bag, clad only in jockey shorts, grabbed a flashlight and stepped out the tent door. The llamas were creating a fuss as I shined my light around the camp and up the hillside. I flashed my light under the lean-to tarp where we kept our food and I saw a big porcupine digging around. I ran to the fire pit, grabbed the ax, intending to chase him away, while yelling for the others to wake up and help me.

Jim says that looking through the moonlight at me with an ax in one hand, a flashlight in the other, almost completely naked and swinging wildly while screaming my head off was the damndest thing he had ever seen. I did kill the porcupine with the ax before he spoiled all our food.

I have replaced my three original llamas with two younger ones and my children pack with them once in a while.

In 1993, my wife Mickey died and camping was out of the question. My own health has begun to fail and I'm certain I'll never return to Willis Lake.

jimmyj 3/25/05

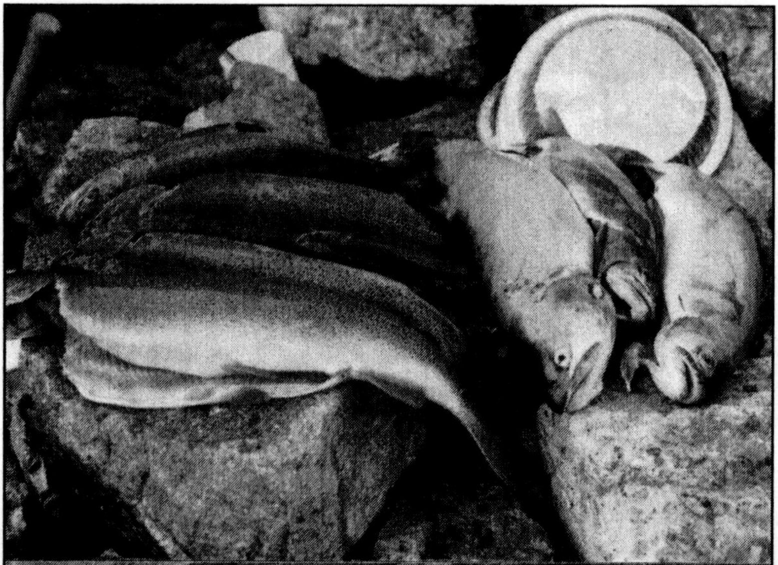

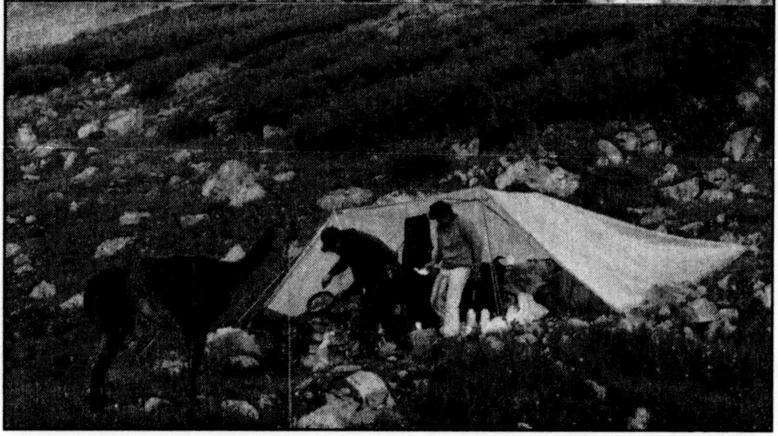

Chapter 7, 1972, Western Operations

As is often the case in the minerals industry, mine operations would vary from an all-out production rush to talk of shut down. The selling price of molybdenum dioxide began to vary widely as other companies began producing moly as a byproduct.

Holing through the Henderson Tunnel

The Henderson project, which was under construction, also spurted, slowed, or rushed with the economy. At Climax, open pit mining began with a jerry-rigged crushing complex assembled out of the old #1 crusher. We spent a lot of time budgeting, re-budgeting and re-budgeting again. As a general rule, there had been more fun each day at the Climax Mine than in a month at Western Operations. In addition to mine operations, exploration, engineering and design, and new mine evaluation were my responsibilities.

Ed Peiker and Tom Cherrier ran Mine Evaluation, Will White ran Western Operations Exploration and Max Gelwix was the Chief Engineer for Climax. So and Will stopped in my office one day in 1976 and said, "Let's make a trip to Canada in mid September," I was hot to go.

It turned out to be quite an experience.

We were allowed to use the Amax Falcon 20 jet for the trip and after the usual delays when going into Canada, we flew into Whitehorse, the capital of the Yukon Territory. The traveling crew consisted of Tom, Will, Ed and I.

The first leg of our trip was to Mactung, a tungsten prospect near Macmillan pass, 350 miles northeast of Whitehorse. We chartered a twin-engine plane, with a destination of an unmanned gravel airstrip next to the Canol Road near the pass. A helicopter was to pick us up and take us to the exploration site.

The Canol Road was built during World War 2 to trace an oil pipeline from Norman Wells in The Northwest Territories to Alaska. You may remember it from an investigation by an unknown senator from Missouri named Truman, as one of the great boondoggles of the war. When completed, the pipeline was too small to carry crude, so an unsuccessful attempt was made to truck crude over the road. Stacks of fifty-five gallon fuel barrels still litter the countryside like so many tin cans.

I cannot recall the reason, but our flight was cut short at Ross River, also on the road, and the helicopter picked us up over a hundred miles from Mactung.

Our pilot was a veteran of the Viet Nam War and he flew as if he was still in combat. Flying with him was not for the faint-hearted. We flew directly into the prospect site and landed on a miniscule helipad built into the side of a mountain. There was a foggy snowstorm at the time and I thought visibility was zero, but you had to have faith or forget it.

The drillers were fed like kings; they often spent weeks at the job. I remember they had flown in ice cream especially for Will White for an evening snack. Breakfast was moose steak and potatoes.

After a review of the latest drill data, we decided to

get an overview of the general area from the chopper to check possible mill and tailing sites.

Caribou were abundant and we saw a small group of Dall bighorn sheep. While flying high over the sparse scrub near tree line, Will thought he saw the sun reflect off some moose antlers. The pilot put the helicopter into a tight spin and dropped several thousand feet in altitude, the rotors on the inside, body swinging around on the outside, with many Gs of force pinning us to our seats as we spiraled down. He leveled out a few hundred feet above some cows and a mad bull moose that wildly shook his horns at us.

We returned to Whitehorse and the next day drove south along the Alcan Highway to Atlin, where we inspected several prospects. At one I was shocked to see a body hanging by its heels in an ore storage bin. You would be surprised how much a skinned 200 pound black bear looks like a human. Ptarmigan were everywhere and seemed as tame as chickens.

The next morning the pilots with the Amax jet were waiting for us at the airport when we drove out from town. We loaded our gear and settled back for the lazy thousand-mile ride to Terrace from where we would fly in a twin engine DeHavallind Otter to Smithers, British Columbia, to see the York-Hardy prospect.

We were climbing sharply and had made a wide turn toward the south, when the jet lurched as we lost power from one engine. There were a few white knuckles holding tightly to the seat arms as the jet staggered under the loss. The pilot made a sharp turn back to Whitehorse and landed safely under the power of the remaining engine.

We waited in the terminal, and after a detailed inspection and multiple conference calls, it was decided to repair the engine with parts flown from the States. That would take several days, depending on aircraft mechanic and parts availability.

Our option was to sit and wait for repairs or charter an Otter to fly us directly to Smithers. Later, the repaired jet would meet us in Terrace. Another chartered Otter would fly from Prince Rupert to pick us up at Smithers and take us back to Terrace. It sounded like a workable scheme, so we flew, somewhat more slowly in a twin-engine prop plane instead of the speedy jet, directly to Smithers.

It was a leisurely visit at the York-Hardy prospect as we waited for word of the Falcon's repair. This prospect had been under consideration for some time and there was extensive tunneling and diamond drilling to review. The deposit was directly under a glacier; I suppose no one will ever know if it is possible to mine in that location.

In a few days the jet was repaired and plans were made to meet in Terrace the next day. The plan was frustrated when the Otter that was located in Prince Rupert and was supposed to pick us up was socked in by the weather. We were sitting in the office, wondering what to do, when a geologist suggested we call Leadfoot Louie,

Louie Dubuc, whom he heard had recently purchased a DeHavallind Beaver and had it parked at Burns Lake.

Ed called him and he said he would be glad to fly us to Terrace. He was to meet us at the little lake that served as the water airport for Smithers.

We had unloaded and carried our gear to the dock when Louie flew in and taxied over to us. We could see the old single-engine Beaver had many miles, but Louie assured us it had just been overhauled. When he opened the cockpit door, we could see there were no seats in back and that the inside had been completely lined with a bright red shag carpet. I believe that was the only insulation it had. He instructed us to put on as many warm clothes as we had and to stack the suitcases and other gear in back. There was barely room for Ed, Will and Tom to crawl in with the gear and brace themselves as best they could for the takeoff. I buckled into the co-pilot seat.

Louie taxied downwind into the weeds as far as he

could before turning into the wind for takeoff. The old Beaver roared and slowly picked up speed as the pontoons lifted in the water. The scrub spruce on the far shoreline appeared to be getting close when Louie tipped the plane enough to raise one pontoon out of the water to decrease the drag. I swear we were in the shoreline weeds when we lifted off, just barely clearing the spruce.

Later, when I told our jet pilot -- a former bush pilot himself -- about that takeoff he said, "You didn't need to worry. That was the safest plane you've used on the whole trip."

The return to Denver was uneventful, thank goodness.

I wonder if "Leadfoot Louie" is still alive?

Western Operations was Amax's shining glory and cash cow. Our CEO, Ian MacGregor was on his way to make Amax "America's greatest natural resources company." Deeply in debt, but beginning to believe their own BS, they embarked on a campaign to let the world know how great they were. I began to pull into a shell, for I could do little to change the business philosophy with which I could not agree.

It became apparent that I had reached the limit of my capabilities, and I was passed over for what should have been an obvious promotion.

The decision to retire came rather suddenly, but not unexpectedly. With my wife, I returned to Platteville to be honored as a "Distinguished Alumnus of the University of Wisconsin, Platteville" during graduation exercises in May of 1982. After a pleasant reception with the university administration, we returned to the motel.

I began to discuss how it never had ever occurred to me that someday I might receive such an honor. Then I mused, "Maybe that is enough. Maybe it is time to move in a new direction, to change my career completely."

My wife said, "I think it is. The stress of recent years is apparent to me. You are losing your hair and turning gray. You used to laugh about your experiences on the job, but not anymore. For ten years you have invested in the nursery, and you are happy when working there. Do you suppose we could earn a living?"

She went on, "You know I don't need much. All the kids are out of school except Chrissy and Melissa, and I believe they would enjoy Buena Vista."

Now the discussion became animated as we considered the pros and cons. Before morning, the decision was firm. We would move back to the farm.

I returned to Golden and after several telephone conferences, it was agreed that I could seek an early retirement from Amax. I believe they were glad to see me go.

I know that I was glad to leave.
jimmyj 3/10/05

The Farm at Buena Vista in 2002.

Chapter 8, In Retrospect

The Climax mine was unique, even among contemporary mines. The underground mine produced as much as 43,500 tons of ore in one day in 1966. When the open pit was operating in 1976, a combined tonnage of 51,133 tons of ore was produced one day. Over a period of 76 years, over 60,000 different employees produced 479,000,000 tons of ore, at an average grade of .33% MoS_2, containing 1.9 billion pounds of elemental molybdenum. There will never be another.

The Climax town, which surrounded the Phillipson portal and the ore concentrating mills, grew to its largest size in the late fifties, with about 1,800 residents. This almost-closed little society began to unravel in the late 1950s and by 1970 was completely gone. The remnants of the Climax community scattered, mostly to Leadville, but also to the surrounding area.

The town infrastructure was slowly disassembled when the houses were sold to employees and physically moved. Employment grew and offsite housing was subsidized in Leadville. The schools were closed in 1962, the hotels torn down and the Fremont Trading Company closed the store, bar, and gas station.

The Climax Mine had a tremendous influence on the towns of Leadville, Buena Vista, Salida and surprisingly, on the entire state of Colorado. Molybdenum production ceased, and should it be resumed, it will be into a completely different political, cultural and economic situation. The mine is now in a state of standby for resumption of production and twenty years is too short a time to really assess the historical consequences of the operation.

One thing is certain. There will never be a return to the days when Climax was a self-contained town, with its own schools, shopping center and entertainment facilities. A town where the Company was the town council and police force, built and maintained all the houses, built the roads and cleared the snow in addition to constructing and operating all industrial facilities that supported the mine.

Another certainty is that whatever the future brings, I personally will not be part of it. I have moved well beyond the Glory Hole, forever.

jimmyj 3/15/05

One event did bring me back to Evergreen and into contact with my old mining acquaintances. That was the death of my old friend and mentor, Terry Fitzsimmons, in October of 2001. His family graciously allowed me to read a eulogy to him at the services at Christ the King parish following a beautiful eulogy by his son, Pat, who said in part:

"There are so many things I want to tell you about my Dad – about Terry Fitzsimmons – his life, his love of life, of family, and of wife. . . his quiet, steady resolve – especially lately, in the face of certain insistent biological facts, facts we all know cannot be reasoned with, cannot be bargained with, and most of all, cannot, and will not, just "go away". He faced those facts the way he faced every other turning point in his life – in a way only Terry Fitzsimmons could, with a single subject-line e-mail, written just three months ago – it read, in an astonishing understatement, "(Not so) pretty good news. . ."

He followed with a wonderful tribute to the man and his family.

I humbly continued with this:

"Do not let your heart be heavy as we gather at this Mass for Terry Fitzsimmons. He died as he had wished, at home with Bobbi and his family. With the family that was his life.

"I can only stand before you as his friend, to speak for those who were his friends. Friends who were unable to help him in his final days, and who can only wish him well in his last journey, friends who are better persons for having known this talented man.

"Terry had the amazing ability to express another man's thoughts in words the man himself was unable to formulate. His skill was so great that listeners were often not aware who had composed the essays delivered by and attributed to others. He was not self-seeking in the use of this skill, he considered it his job, and contributed it willingly and forcefully.

He was the ultimate public relations man.

"It was the use of this skill that brought him to the Climax Western Operations in Golden in the early 1970s. Leaving Leadville, the town of his birth, and the cozy confinement of the Climax Mine, he further developed his skills until he became the voice, the spokesman for Amax in the West. Often it was Terry who stepped forward to field the probing questions of a critical media in situations that left most of us confused and inept.

"During this time, I had the distinct pleasure of becoming a close personal friend of Terry's as we raised our families together in Evergreen. We built the deck on his house together, hiked to Willis Lake together and stayed together at Gene Bond's cabin, the now historic site on the south shore of Twin Lakes. Many an interesting conversation developed over a cold brew with this sometimes not quite Americanized Irishman, as we both tried to discern the meaning of life.

"About fifteen years ago in this very church I listened to the eulogy at another funeral, the funeral of another Terry Fitzsimmons, a son in this fine family. The priest who delivered the eulogy remarked that God must have had a project consisting of much hard work that needed to be done. Why else would he have called one so skilled in the way of work, so clearly before his time?

"Today in a most torturous of times our God is being claimed by each of many cultures and their many factions. Man's arrogant attempt to define God to his own perception has led to misunderstanding, conflict and terrorism. Would it be presumptuous to suppose that God has again felt the need to ask another Terry Fitzsimmons to use his skill with words to assist him in defining for all people his true nature, thereby bringing peace and understanding to a troubled world? Perhaps it is, but I can assure you that if asked, Terry will do his very best. He simply knows no other way.

"May he be Blessed for all eternity."
Jim Ludwig,
October 2001

Chapter 9, Reflections

When our family returned to Buena Vista in 1982, I deliberately distanced myself from the Amax operations and the mine at Climax. A small bit of consulting made me realize it would be best to move on, and I immersed myself into the operation of our nursery.

This was the first time I began to reflect on my life, my family and experiences I had in the previous fifty years. I realized that I knew very little about my parents, my siblings and relatives.

I became determined that, unlike myself, my family would know who I was and where I came from. The advent of computers allowed me to overcome my lack of communication skills, and I began writing Christmas letters and other stories to my children.

Reflections on Growing Old

There was a time when I could throw a ball; I mean really throw a ball!

I'll never forget the day Hi Gordon in the announcer's booth called out, "The count is two and two, here comes the pitch, and he hit him again!"

The center fielder, never one to mince his words, said, "Jim, you're a poor man's Ryne Duran." (A Yankee fastballer those days, who wore thick eyeglasses.) "When you stare down at the batter with those beady eyes through those coke bottle bottoms that you wear, the poor kid knows you can't see him, much less the plate your aiming for. That last pitch wasn't much inside, but he froze like a pika dead with fright."

I didn't pitch much after that, another skill I really never had, lost at the age of twenty-six. I was already getting old.

Then one day ten years later, when the neighbor's cat was stalking birds at my bird feeder, I picked up a rock and threw my arm forever out of joint. Now, the Doc says it is arthritis; take an aspirin when it aches. Jamie, the masseuse, works it over now and then, but it aches anyway. Now I know I'm getting old.

There was a time when I could walk; I mean really walk!

With a pack upon my back, a fishing pole in hand, I've walked the Gore and Sawatch, seeking lakes with native trout and seeing views there were no other way to see.

When my fears of being packed out in a body bag became too strong, they bought me llamas to take on the load. That added ten more years of walking to the highest lakes. In 1993 I quit. It wasn't fun to walk and hurt and not enjoy the beauty of the High Country.

Now I walk a bit to keep my circulation up and ride the four-wheeler around the farm.

Now I know my poor bod is old.

There was a time when I was strong; I mean really strong!

I'd pack a rock drill through the stopes on slopes of forty-five degrees. The damn thing weighed a hundred pounds or more. I could throw a ten-pound bag of powder fifty feet or so, set timber eight by eights over my head.

Now the words that I most often say, "Will you give me a hand?"

The words I often hear, "Don't hurt yourself. I'll help you, Dad."

I can tell I'm getting weak and old.

There was a time when I could dance; I mean really dance!

Three polkas in a row would warm me up, another beer, and then dance some more. Hoot and holler, stamp my feet, clap my hands and round and round again.

But that was years ago, now half a polka song will send me to my seat. My heart will pound, my lungs will hurt, and my legs will ache. I'll go sit down and hold her hand and tap my feet.

There was a time a pretty girl would interest me; I mean really interest me!

I saw one walk by the other day. I knew I should be interested; could not remember why.

Did you notice how each verse got shorter as I wrote? I mean really short!

That's the way it is when you grow old.

I guess I should feel badly.

But I don't.

jimmyj 11/12/97

How to Choose a Cantaloupe

I was startled from my noontime nap!

"Honey, run to the grocery store and pick me out some fruit. I need to make a salad when the girls come to play cards. I promise to love you forever if you do. Don't spend too much money or bring me home some garbage like you sometimes do."

Grumbling to myself, I walked into the grocery store. Why can't she let me snooze a little more? An old man needs his rest, you know.

I took a basket, eased my way, and looked up at the sign. "PRODUCE NOW ON SALE." Let's see now, kiwi, two for $0.99, Navel oranges, $1.79 a pound. Hard green Granny Smiths, only $0.69 each. Bananas, pay a fortune for a bunch and get one free.

And then I saw the cantaloupe direct from Rocky Ford, at only $0.19 a pound. That's what I need!

They were stacked row upon row, filled an area six by eight. Customers were picking through the stack. I nonchalantly watched to see exactly how they picked one out, because I remembered then the warning of my wife, to not bring home some garbage like an old man tends to do.

I decided I would watch some more; turned up my hearing aid. Perhaps I could learn a tip or two from other customers as they pawed through the stock. I might even ask some questions for advice, something a man will seldom do.

A young mother with two kids in tow would rap each melon with her knuckles, listen intently to the sound. I tried to hear. The sixth one that she rapped, she rapped again, then smiled in satisfaction, gave it to the urchin in the basket to take home.

A matron, who looked like she really knew, would press the spot where the stem had been, then squeeze a bit and smell each fruit. It only took her five prospective choices, until she went back to the first she had examined, and chose it.

At last I saw a neighbor I knew. I watched as she examined each so carefully, rapped and shook, squeezed and smelled. I asked her if she wouldn't mind to take a minute and explain to this old man the way to choose a cantaloupe.

She said, "I'm really sorry, Jim, but I don't have a clue. I only rap and shake and squeeze and smell, so other folks don't think I'm stupid when they watch me choose a cantaloupe."

Here comes a stock boy with more cartons from Rocky Ford. Surely he will know. He said, "Look at a any one inch square, exactly one hundred degrees from the stem end. The lace should be an even tan, the background slightly green and firm to pressure from your thumb. It should have, you know, the smell of cantaloupe, not too strong and not too ripe, but smell it must. The seeds should rattle just a bit when shaken hard. Like this!"

I decided he was pulling my leg and really hoped that I would pick up garbage to clear the rack.

A well-dressed businessman, his basket stacked high, picked two from in the middle, one from each corner of the pile. As he rushed away, I asked him, "Pardon me, sir. It is obvious you know the way to pick a cantaloupe. Do you always pick them right?"

"Never miss," he said. "I take them home and cut them open with a knife. Then I throw away the ones that stink, or are too green to eat. At nineteen cents a pound, I couldn't care less. I drop forty bucks on the lottery every week, and have never won a dime. A two-bit cantaloupe won't break the bank."

I put two in my basket, which looked as if they might fall from the pile onto the floor. The checkout lady smiled at me and said, "You sure know how to pick a cantaloupe. These two are simply excellent. I'm sure your wife will be most pleased to have a husband with such skill."

And pleased she was. The melon balls she made were sweet and soft. She said, "Oh, thank you, Honey. Tell me how you pick a melon like you do."

So I said, "First you pick a one inch square, exactly one hundred degrees from the stem end. The lace should be an even tan, the background slightly green and firm to pressure from your thumb. It should have, you know, the smell of cantaloupe, not too strong and not too ripe, but smell it must......"

jimmyj 9/7/98

Trucks and Things

Have you ever seen my daughter Linda bowl? I haven't for a long time, but the last time I did she held the ball on her left hand, with the fingers of her right hand in the holes and when she swung the ball, her elbow had a sort of crook to it, sort of like left hander Denny Naegle throwing a curve ball with his right hand. I didn't have to ask her why; I knew. It is not what most people think -- that Linda just does things differently. It was because she fell off the truck.

Well, not exactly, because it would have been OK if she had just exercised properly after she fell off the truck and broke her elbow. She just kept running around with her elbow in a crook, holding most anything in that hand as an excuse to keep from straightening her arm. Now she is stuck with a 175-pin average, and I think that heavy ball has straightened her arm.

You might ask, what truck was that? Surely she shouldn't have fallen out of a truck box?

Well, it wasn't the box; it was off the top of the cab. And what was she doing on the top of the cab? Heaven only knows, and it is not too sure.

Anyway, it was just a tiny truck, that old flathead '48 Jeep I bought from Shorty Lester. Shorty sold it to me for $150 because it was too small to pull his horse trailer. At that, it was a better bargain than the Shetland pony he later gave me.

I really loved that truck, our very first one. We really couldn't afford it, but I had all kinds of excuses. I needed it to haul trash, especially since we had a big garden. And I hoped some day to buy the Jordan house, and I would have to haul wood for the fireplace. (You have all heard Dad use this logical reasoning to get his way.) But most importantly, I needed a four-wheel drive to get around the Four-Mile Ranch with the kids.

Now that I've thought of it, let me tell you about the pony. Shorty bought a $1.00 ticket in a raffle at some small town celebration and he won a Shetland pony. When he put it in his trailer, the guy said he might have trouble keeping it in a fence. Shorty called me to see if he might keep it at Grandpa's place in Twin Lakes, then kids could ride it if they wished.

What a deal, a pony for nothing.

So he took it to Grandpa's, opened the gate and drove into the field. He let the pony out of the trailer and by the time he had driven out of the field, the pony had already crawled through the fence just down from the gate. He had to chase it back in several times before it would stay there. He said he had to beat it a bit to make it behave. Shorty was not a horse whisperer!

The kids could not wait to ride the pony, so we drove up to Grandpa's to give it a try. That was the meanest-looking animal I had ever seen. Coal black, his back was lower than my belt. He had shifty eyes and ears that lay back if you spoke to him. I had a halter on him with a lead rope. He stood quietly until Mark sat on his back, then immediately dumped him in the bushes. I remembered how Shorty had to "beat him a bit" to keep him in the fence. I set Mark back on and when he went into the bushes again I whipped him with a willow branch. I mean whipped him. Then he let Mark ride and stayed clear of the brush, but the kids decided not to ride any more.

I called Shorty to come get him. He gave his raffle prize back -- just dumped him out of the trailer because the former owner didn't want him. For some reason, none of you kids ever begged for a pony.

Anyway, it must have been 1960 or 61, and we had a truck. The driver's door would not latch and the window wouldn't roll up. It was a four-wheel drive, but it was such a battle to lock the font hubs that we seldom used them. It took a wrench, and then I had to rock it back and forth until they dropped in. It would not stay in second gear, even if I held the shift lever, so it was either low or high, and I needed just the right combination of backup and clutch slipping to put it in low range. It had a leaky brake line somewhere, so I always had to be ready to grab the hand- pull emergency brake lever near my left foot. Sometimes that was hard because my left arm would be out the open window holding the door shut, and my right hand was holding it in gear, so my chest was squeezed against the steering wheel, keeping it on the road as long as that was straight ahead.

Mostly it was just parked in the back yard so the kids could climb in and out, up and over, and fall off the cab.

Ray Sailor would let us drive around the Four-Mile Ranch, so we would picnic by that big cottonwood clump. We would look for arrowheads and one time found a chert hide scraper and an awl for punching leather. Once there was a dead bobcat in that little cabin up in the piñons. I threw it on a big anthill so they would clean off the skeleton. Somebody else found it, I guess, because suddenly it was gone.

At first there was nothing in the old house, so we would snoop around there and in the barn. Later Ray and Inez fixed up the cabin so they could stay in it while working the cows.

One time we saw a bighorn ewe with the cows. When we tried to get close, it would get up, run a hundred yards, then lay down again. It was so obviously sick or injured that I went to see Cody Jordan, the game warden, and asked him to look at it. When he didn't call me back, I approached him in the old hotel bar and asked him about it. He told me he drove up there, but the big Husky that always rode in the back of his warden's truck jumped out and killed the ewe before he could stop it. He didn't even look to see what was wrong with the sheep.

That was Kedran (Cookie) Jordan's dad. She used to baby-sit our kids and we liked her name so well we gave the name to our next baby girl.

We took that old truck fishing to Clear Creek Reservoir and would park along the highway, then walk through the woods to the south shore. There was always a nice beach over there where the kids could play in the sand instead of fish.

Sometime after we moved into the Jordan house, I traded the jeep for a big red three-quarter-ton Chevy with a stock rack. I made some hoops and bought a canvas to go over the top of the stock rack so we could sleep in it.

I did that because my wife Mickey would not sleep in a tent. We camped several times at the creek in Four-Mile. The kids would sleep in that big old green canvas tent Charlie Jordan had given to me, and Mickey and I slept in the back of the truck on a mattress. Camping was not her thing, so soon she would not consider a camp-out at all. She said her idea of roughing it was staying at a good quality Holiday Inn.

Chevrolet finally made a four-wheel drive truck and I traded for a short, bright blue one we used to take ice fishing on Clear Creek and salmon snagging at Twin Lakes.

Mechanically, that was the worst vehicle I ever owned. It had a three-speed steering wheel shift and no low range. I remember thrashing around in Lake Creek with the truck like a mad bull in a china closet after breaking through the ice.

When we made the first road through Dr. Lipscomb's, I had to roar through the creek and up the bank to keep from stalling in the lowest gear. Mark said I was completely airborne as I came over the top. That road became the Bureau of Reclamation's main road to the south shore and is still in use today.

I sold that truck to Frank Thomas and bought the best truck we ever owned, a gray three quarter ton Chevy. It was our main nursery truck for years. It hauled many a llama to the Willis Lake trail. That one is still in service here in town as of 2003.

I really was attached to that old Jeep, but after that, a truck was just a better way to haul things.

Maybe if I bought a Lincoln or Mercedes town car, which I can't afford, it would become more precious to me. I could wash it and vacuum it and polish it and park it at the nursery so everyone could see what an expensive vehicle I have. Then one of these years I could buy grandson Sam a chauffeur's hat and uniform and he could drive us around to all the big shows, concerts and operas.

On second thought, I think not. I think that God is warning me not to drive at all, or the safety of all other people. He knows I can't think fast enough, or react fast enough, or keep my attention on the road, even if I could see where I'm going. He knows I might fall asleep at any time even with my eyes wide open.

Maybe I had better listen to Him.
jimmyj 3/9/2003

On Fixing Things

When I was just a little lad I had a toy -- a metal truck -- red in color and my pride and joy. I'd build my highways in the dirt, drive around and haul some sand and just pretend I'd grown up. Until one day a wheel came off. I guess I pushed too hard.

I went to Mom, "Can you fix this?" I asked, as tears welled in my eyes.

She said, "Of course I can, although I won't. Fix it yourself, you're big enough."

I dried my eyes and took the truck out to the shop where Pa was always fixing things. And sure enough, I fixed it. That started off a life long task of fixing things. A task I thoroughly enjoy.

You may have noticed that the first thing I did was break the truck. It seemed that many things I used were broken when I quit, and then were set aside to be fixed. In the meantime I saved all the pieces, and found more that might just work for fixing things. When I got around to fixing the first thing I broke, I had lost all the pieces that I had saved to fix it with.

My life has been a contest, a four-way one at that. It seems that I am either:

breaking things,
fixing things,
saving things, or
losing things.

Now, breaking things takes little skill. My clumsiness and poor eyesight easily combine to break most anything. But fixing things takes tools, and I spent many hours on scheming to buy this or to buy that, because I needed just this one for fixing things. Unfortunately, each purchase upped the chance of losing things, or breaking things and added one more chance to saving things.

When I was a boy, breaking things was way ahead, fixing things improved, saving things was gaining, and I lost almost none at all.

As a young man with family, fixing things was all I did. At least in between saving things, and breaking things while fixing things.

I even went to welding school, and then bought the tools to cut and grind and drill and weld, until I could fix anything. At least I thought I could.

My tool chest then had many things, and shops and garages were half full of things saved just for fixing things. So then, to keep from losing things, I built more garages and made basements large to stash my loot.

As I grew older and then retired to the farm, we had more buildings, green houses, tractors, trailers, tillers, trucks, and all the hands to run them, and to break them down for me to fix.

Fixing things took most of my time, except that I started losing things that I was saving just for fixing things. I had no time for breaking thing. I spent my spare time finding things that I had lost, and that I needed just for fixing things.

Now I'm older still and have slacked off on fixing things. My eyes don't work as well, arthritis makes my fingers stiff and I'm so clumsy that a bib should be required just to catch the soup that dribbles from my spoon.

I'm excellent at breaking things, and saving things has slipped way back, but where I really now excel is at losing things. I leave the phone off of its rack, the marker in a book is lost, my checkbook is missing, and the coffee cup remains behind the big reclining chair. Papers have been filed; I don't know where. The four-wheeler is parked right where I drove it, then left it and walked back. My jack knife was right here, I know, but where?

I've found a way to even out the game by giving things away. If I can't fix that broken thing, or find the tool I need, the part I had, I must have given it to one who wanted it. Now I don't give so much away, but just enough that if you ask for something I lost, then I will say, "I must have given that away."

Everyone has been most kind. They help to find the things I lose, they fix the things I break, and take away the things I saved.

I spent my life fixing things; no man could ask for more.

jimmyj 2/5/97

On Watching Birds

The early memories of life are those of living on the farm. We may have been poor, but only as poor as we allowed ourselves to be. In one way we were rich -- our contact with the great outdoors -- and in the outdoors there were birds. I watched them then, I watch them now.

The crows came north in March They would hang around the barnyard and with their raucous call, let the world know that spring was close at hand.

The sparrows stayed all winter, trying every trick they knew to get into the granary and eat the oats stored there for cattle feed. We kids in turn tried every trick to shoot them with a BB gun, without shooting through the barn or granary roof.

With spring came bluebirds, robins, meadowlarks and bobolinks. The partridge boomed from strutting grounds back in the woods, the nighthawk's wings resounded in their mating dives. The killdeer called, and led us from their nest with pseudo broken wing.

When I was young and curious, watching birds taught me patience and filled my mind. It didn't cost a thing to watch the birds.

Some other lessons that I learned about birds were harsh. I remember the first chicken that I saw my mother kill by chopping off its head. The life drained from its eyes, the headless body lay on the ground, the wings flapping wildly as the blood drained from the neck.

I remember baby chicks. The cuddly things would pick another chick to death. I knew more of life's realities at five than kids today know in their teens.

At the winter lumber camp in '35, we put the table scraps out on the cook shack windowsill for the birds. It was here my mother taught me to appreciate them, to note the subtle differences that made each kind unique.

For instance, Nuthatches circle around a tree trunk going up, Brown Creepers circle around while headed down. A Chickadee will sing and eat while hanging upside down, holding a seed with its foot to crack the shell. Bluejays scream and steal from other birds, fly off and hide their loot, and then come back and steal some more.

I think woodpeckers peck to hear the noise they make, they circle around and hide behind the trunk when approached. The Gray Jays sneak so silently, cat burglars floating by on muffled wings. Camp robbers they are called.

"Watch the way they act, observe the way they fly," my mother said. "Then you will know a bird from far away. See why each is different and remember it that way."

When I went back to school the winter of '35, I was hooked -- a bird watcher for life. I had learned without effort, the scientific method to research and learn the unknown.

I watched and learned through all my boyhood years: how the swallows built their nests of mud under the eaves; listened to the thrashers and the cat birds imitate the meadowlarks and bobolinks; watched the bitterns hide among the cattails by pointing their long bills straight up.

The only geese we saw were flying high in a long and wavering "V", headed north to Canada. We hunted ruffed grouse, "patridge" to Pa, and once he shot some prairie chickens in the pasture back of Gebert's farm.

When I moved west a *"Field Guide to the Western Birds"* was the first book I purchased. I've never been without one since, and have worn out a few.

I've seen ptarmigan sit in the snow under the clotheslines at our house in Climax, hunted blue grouse high on Copper Mountain, long before it ever felt a ski. I've watched ravens float on updrafts of the north wind swirling in the Glory Hole. Pine grosbeaks, white crowned sparrows, gray crowned rosy finches nested in the alpine fir and krumholz of the timberline. Water Ouzels dipping in the mountain streams were quite a sight to see.

The press of life, the stress of work could be forgotten in a moment with my binocs on a bird.

Soon we had kids to teach to watch the birds. We started feeding winter birds in Bueny at the Jordan house. I built a platform between trees next to the garage. The piñon jays came squawking by in flocks of twenty-five or more, the Stellar's jay was now our local thief. The evening grosbeaks came in flocks to strip berries from trees and gray-crowned rosy finches swarmed our feeder when the winter storms would push them down from the high slopes.

A Clark's nutcracker showed off his formal suit of black and gray and white, and pecked on frozen suet wired to a limb, scaring off the others that were there.

I remember when a cattle egret with a damaged wing appeared from out of nowhere on our lawn. Mark and I crawled on our hands and knees, covered up with raincoats in the rain, just to get a closer look. A passing car slid to a stop, the driver stared and shook his head.

In Evergreen we had a feeder just beyond the deck outside the northern door. It was so close, we could identify the tiniest of birds. There were White, pygmy, and red breasted nuthatches, finches of the various kinds, mountain and black capped chickadees, starlings and English sparrows. Stellar's jays, camp robbers,(Canada Jays), a titmouse and a cedar waxwing were a rarity. An Abert's squirrel with tufted ears would steal the food.

Migrating crossbills in their fiery coats of red would brighten up the spring. We didn't see them often, for when the spring arrived, every weekend we would drive to Bueny and spend the weekend working on the farm. The water would come down the ditch by April 1st and spread across the meadow, bringing back to life the meadow grass and all the various bugs and worms that called the meadow home.

The Franklin's gulls, perhaps a ringbilled gull or two, would spot the water from on high, with raucous call the flock would land, and like a pile of leaves caught in a devil wind, would swirl and rise, then settle back again. A flock of ibis, red eyed, black and nearly as large as turkeys, would punch their long, curved bills into the swampy earth, extracting earthworms for their food. The common snipes would burst from underfoot with whistling wings.

Flocks of blackbirds, Brewer's, red wings, and yellow headed, would sing the songs of spring from perches in the cottonwoods along the lane. Pairs of geese and mallard ducks and on occasion cattle egrets or a willet would be seen.

It was a birders heaven.

When I retired we moved back home, back to the nursery we had started just ten years before. We set our summer's clock by the springtime migration of ibises and Franklin's gulls. One year Audubon's warblers stopped by, I've only seen them once in forty years.

When the summer came, we watched the nesting, had a half-a-dozen houses every year. There were mountain bluebirds, violet-green and tree swallows all harassed by the English sparrows trying to take away the nesting box. Barn swallows would build their nests of mud on rafters in the llama barn. We had a female bluebird that roosted above the front porch light for five years in a row. She seemed to find a new mate every year and took her choice of all the boxes in our yard.

Each summer there were half a dozen brewer's blackbirds nesting around the nursery. One year a white-crowned sparrow nested in a gallon shrub. I saw a bobolink last summer, the first I've seen since I moved west.

Last winter was a strange one, some robins never left. A Townsend's solitaire stayed all winter in a juniper outside the Nursery door, and meadowlarks were on the lane in March. I added a Say's phoebe to my list on April first.

I've often wondered why the birds fascinate me. I never can remember TV programs and jokes and stories slip my mind. But ask me of a trip I took and I can tell you that I saw a heron on the lake, or heard the laughter of a loon, and watched him dive and surface with a fish. The only memory of one business trip, I stopped the car and watched some western grebes do their dance side by side across the water.

My binocs and bird book are always ready by the sunroom door. From there I can watch the feed and water we placed on the deck to entice the birds that I so dearly love. The birds have helped to make a life worth living for this rather odd old man.

Some day when I no longer walk, they'll wheel me out to see the gardens they have made for me. I'll ask for my binocs. I'll need to have them just in case there is a bird that I have never seen.

I also better have them when they send me on my way -- I've never seen a pure white dove.

jimmyj 4/5/97

Addendum: On Watching Birds

It was three p.m. on April 25 of 97. It had snowed for a day or so and most of the snow had melted as it hit the ground. I had ridden down the lane, upon my trusty steed, a Honda with four wheels. Out in the fields, eight inches more or less, had covered all the grass, with just a bare spot where the irrigation water flowed. In that patch of open grass there was a flock of birds.

Two pairs of greater Canada geese strode regally about, made a dozen Ibis move out of their way. A pair of mallards wallowed in the muddy meadow and ignored the flock of Franklin's gulls. A couple of Ring-billed gulls tried hard to find something to steal. A pair of Ravens watched from in the snow as smaller birds, a bunch of robins, a couple western meadowlarks, a pair of killdeers, starlings and some blackbirds, scrambled to stay out from under foot.

What a sight, a Mother Nature circus, free of charge.

It was a little patch of green, a haven for the birds when winter made a short return to Pleasant Avenue, our home in the Banana Belt of Colorado. By Sunday the snow was a memory. The birds all dispersed, some migrated further north, and we waited another year, for the circus of the birds.

jimmyj 4/25/97

The Garden

When I was just a little lad, my mother raised a garden. At least it seemed like it was hers.

I know Pa hauled manure and plowed, and on a summer's eve, he would walk around and taste a pea, pick a radish, or pull a weed. It was a long day on the farm, and now the work was done. It didn't dawn on me that he loved the garden. He relaxed while there and for a moment all his cares were gone.

It was a chore for Mother. She had to seed and see that little plants were started in a hot bed that was built to shorten the Wisconsin winter just enough to make tomatoes ripen or a crop of cabbage grow for the sauerkraut and canned goods that would fill the cellar by the fall.

She had to keep it weeded, which meant trying to get kids to share the load and earn their keep.

I was undisciplined and lazy, as I look back now. It may have been much easier to weed the place herself, which she often did. A garden was a part of growing up; I hated it. How often have I heard my mother say, "You can't go playing ball until you've weeded all the beans, and not a minute before then."

How I wished that they had a spray to kill the weeds, or a spray to kill the potato bugs we had to pick by hand and drop into a can of gasoline.

The year that I was twelve, I was already working out, away from home in summer, just for room and board, and a couple of bucks a week.

At least I didn't have to weed the garden then, and didn't for a lot of years. But in the winter, I still feasted on the things my mother grew and canned.

In ten more years a mother-in-law would take her place -- my wife would cook the produce from her yard.

I watched my dad and Frank, my father-in-law. They seeded, hoed and harvested, and as they puttered about, I realized this really was a work of love.

When I was thirty, more or less, and had a wife and family, we moved down from the hill, away from Climax town, from where the springtime's poppies bloomed in

August, and columbines would push their blooms from underneath the snow.

We moved to Buena Vista, to an altitude of about 8000 feet. It was a welcome change, although we still might have a frost in June and a frost on Labor Day was not unknown.

Then I began to garden with a vengeance, as if by working hard, I could regain the knowledge of the soil that I had lost in the eighteen years since I had left the farm.

It doesn't work that way. The seeds will teach you patience, and the dirt will humble the most arrogant of men.

The seeds accumulate the warmth; assimilate the moisture over just the length of time, and in just the right amount that God intends. Then, and only then, if He is willing, a plant will grow.

So it was with learning how to garden. Take a little bit of this, try a little bit of that, then watch and try again. Remember what I did, and maybe, if He wills, the next time it will grow just like I want it to.

As years went by, the garden was a part of life. Each year a new variety would prove itself. A flower, then one more, and then another, would find the perfect place in our big yard.

Just like my dad, I grew to love the garden and can't imagine living somewhere where there isn't one, or where a garden will not grow.

A journey to Wisconsin would never be complete without a careful look at Grandpa Louie's garden. He sometimes was alone. He outlived three wives, and always had a garden, although he gave away the most of what he grew.

One year on my vacation when he was seventy-six, I helped him plant an apple tree. "Just so he'd have some fruit when he grew old." He lived another fifteen years, ate many apples from that tree.

Grandpa's Garden

    And so with Grandpa Frank, who gardened when he could not walk, and needed help to get back in his wheel chair. He grew gladiolus in profusion, and gave bouquets to the church and anyone less fortunate than he. The last years that he lived, I planted a garden for him and tended it as best I could. We pushed his wheel chair on the lawn; the breeze would softly move the leaves as he sat and watched his garden grow, planning for another year.

    Now as I grow old, I find my love of gardening to be as strong as theirs. It's wintertime, the first of March; my garden patch is covered with new snow. But in my pocket on this day are packs of seeds. I bought them to grow hotbed plants -- plants like my mother grew more than sixty years ago. I'll do it every spring, for that is how to start. It is the easy thing to do. Then God takes charge and does the work and I just stand around, leaning on a hoe, and watch his garden grow.

                         jimmyj 3/1/97

Chapter 10, An Old Man Learns to Weave

We had three llamas: A white one, Frolic; Thistle with the chocolate wool; and Cider, part quanaco, he was throwback to the wild ones, with pointed ears, a shining breast of white, a cider-colored back and on his face, a blaze of white, no wild trait, it should be gray.

When I would brush them clean and blow the chaff and dirt away, then I would gather in my hands the llama wool for all to see. It was so light, so smooth, so soft, so beautiful and dry without the lanolin that makes a sheep's wool smell and feels like grit upon a dust cloth in your hand.

I wondered how the yarn would be, made from this fiber, for I knew the Incas dressed most lavishly in cloth made from vicuna wool, a distant cousin of my three.

Then in a western book I saw a Hopi maiden spinning wool upon a spindle, held in hand and twirled twixt her fingers. The wool would wrap around the shaft, she'd make a ball and then return to place the fiber in her loom. A rug of beauty would result from simple things applied with skill.

I can do that, I thought. I will!

I didn't know I had started on a task that for the next six years would use the idle hours of an old man in retirement, until it grew to take all the time and emotion that an old man has to spare. I found some wood -- mahogany, that was tight of grain and so heavy that when a wheel delicately fitted to a shaft of oak, then with a string suspended high and spun with speed, digitally, would turn and turn 'til stopped by me.

I took a bunch of llama wool and tied it to a string that I had tied on to the spindle shaft and with a half hitch fastened it onto the spindle tip into a notch that I had whittled there. Then with my fingers gave a twist and much to my bewilderment, the yarn began to form and pull the silky fibers from my palm until a length a yard or more extended down to near the floor.

I stopped the spindle in its flight, and wrapped the

length of yarn about the shaft; half hitched the tip and spun again. Another length of yarn appeared, for God and I had given birth to yarn just as some ancient one millenniums ago had done to supplement his furs with cloth against the winter cold.

My skill increased, but many times the floss would not draw smoothly from my hand. I bought a book on Navajos and saw how they would card the wool and make a roll from which the wool came easily, and made a yarn as soft as silk.

I wandered to an antique store and found a pair of cards that some pioneer women carried west from in an eastern outfit store. A worn wooden board, just four by six, that had a handle mounted at angle on the back. A piece of leather was tightly stretched across the face from which protruded tiny wires of hard spring steel. Each was spaced precisely from the last and each was crooked one- fourth an inch, then all were cut to the right length.

In a Foxfire book I saw a woman from the hills of Tennessee carding wool upon her knee. I learned to do this, yes I did.

I placed a card upon its back and held the handle from the left. Upon the card I placed some wool, about so much, not any more. I held the other card, its handle to the right, and drew it firmly cross the wool that's on the card in my left hand. The wool will straighten with the draw, and transfer to the right hand card. I reversed the cards and drew again. The wool transferred much more easily. I reversed again and drew again, then doffed it off into a roll by combing backward on the card.

Now in my hand, a length of fluff. I started at the end and with a twist I tied it to the piece of yarn that's on my spindle and spun again.

My yarn was smooth, and soon I had a ball of twenty yards or more.

One night in winter as we sat before our fireplace and listened to the wind that moaned and groaned, and slanted snow across the yard, my spindle hummed, a peaceful scene. I wondered what had happened to the spinning wheel my Grandpa had when I was young -- a tad of only five or so -- and he was old, a little man, his hair full and gray, and a mustache trimmed beneath his nose.

He'd sit and spin without a word, while Grandma knit the mittens that she gave to us at Christmastime. I think of him no other way.

I called my sister, Betty, the next older one from me, for she had helped remove the things my father had when he passed on. And from the attic, if perchance the spinning wheels had found their way to some one else, some other place, Betty would know.

"Why Jim, I have them both. They're in my basement covered up with an old quilt that also came from Grandma's house. Remember how they used to sit and spin in their old house, by the big stove that kept them warm in wintertime? Your Pa has given one to me, one to Lucille who hasn't room to keep such junk, for they are in such bad repair they're worthless now."

"May I have one?" I asked her, eagerly.

"Oh, take them both. When you return in summertime we'll take them all apart and pack them in a box for you to take on home with you."

It was winter when I looked into the box of spinning parts. I separated out that which appeared to be the best of each and chose to build again the smaller one made out of oak.

The spokes were loose; I took apart the rim and shortened it a bit so I could pull the spokes in tight. The flyer and the bobbin both were broken, and the flyer hooks were gone.

I made a pattern and replaced the homemade shaft. The wool had worn a grove so sharp that llama wool -- so soft and light -- would never pass to twist up tight.

I glued and sanded, clamped and planed, then checked the balance of the wheel.

From nylon string, I made a belt; on the third try the length was right. At last I had a spinning wheel just like my Grandpa used some sixty years ago.

And now to learn to run the thing. I carded wool, a beer box full. I took a piece of yarn and tied it to the spindle of the bobbin, through the flyer hooks and out the shaft. I frayed the end and then attached some fiber from my roll in hand.

My other hand was on the wheel, my foot on pedal down below. I gave a gentle push and pumped, the wheel made half a turn just right then changed direction of its flight. I couldn't stop, I pumped again, the wool came loose and tangled in the flyer hooks. I dropped the roll on to the floor, I pumped again, another time, it changed direction back again. I tried again with less success, I tried again, now practice, practice, practice.

No matter how practiced, I didn't have that same smooth rhythm Grandpa used to have.

I found a motor on an old rock tumbler, mounted it upon a board, made a belt of nylon string and set it so that I could tension with my toe and make the wheel smoothly turn. Soon I was making yarn like Grandpa used to make. My yarn improved in quality and I had to show it off to all my friends.

Then someone asked, "What will you do with all your lovely llama yarn?"

"My mother taught me once to knit or just perhaps, I'll build a loom and learn to weave," I said.

Another said, "An elder uncle has passed on. He had a loom that no one wants. It's not too big, just twenty inch and broken down. But you can have it if you wish."

When I picked up the cast-off loom, I'd never seen a loom before. I didn't know just what was broken, didn't know the names of things scattered about yet obviously a part of this same loom.

So I went out and bought a book, *The Key to Weaving*, by Ms. Black. I read and checked and learned the names, and from a nameplate, found a place in old Quebec that still had parts to fix this thing.

Again I glued and sanded, clamped and planed.

The reed, a comb-like thing, which keeps the warp strings spaced, was bent and damaged. It had to be replaced, then fastened in the beater.

There were so many words that I had never heard: "The harness frames are tied to heddle horses at the top, and at the bottom to the lams, which are fastened to the inside of the upright beam and to the treadles." There were ratchet wheels, dogs and warp beams, pawls and cloth beams. There were sheds and shuttles, swords and shed sticks.

Slowly each piece fell in place, and then I had a loom: a twenty inch, free standing, four harness with rising shed, not quite like new, but it would work for an old man to learn to weave.

I looked into the book again. Just sixteen simple steps to dress or place the warp thread on the loom.

But I first must make a warp board. I was amazed to think that ancient ones had placed some pegs into some holes, and using them, had learned to keep a thousand strings untangled, so they could then be used to dress a loom.

My warp board was quite simple, some half-inch dowels beat into holes drilled in a two by ten. With it I could make a warp of fifteen feet and tie the cross just right. At eight threads to the inch, and eighteen inches wide, 144, plus two extras on each side, I didn't really think that I could keep the threads all straight and could I really place them on a loom? By now I'd come too far to quit; I must go on.

Carefully I tied a bright red guide string to the peg marked "A," and traced the pattern round the pegs with bright red yarn. Then I tied the warp yarn to the same peg "A" and followed the bright pattern until "M" was reached, then back again to "A," making sure a cross was always made between the pegs of "B" and "C."

After each eight warps, I would softly tie the eight together, to help me with the count and keep the warp untangled. I added warp on warp until all 148 were on the board, and neatly tied in groups of eight.

The skill here was in tensioning each warp as it lay on, so when the warp was loose, each string would be equal length. I soundly tied the lease, or porrey-cross with heavy string, then took three groups of eight and tied them loosely on around the board to have ease in handling all the string when each was fifteen feet in length.

I removed peg "M" to get some slack, then chained the warp into soft loops with an old crocheting move, using right hand as the hook, lifting loops of warp with left, and soon a pile of warp lay on the floor.

Now any weaver reading this can see I have already made an error, as I really did the first time that I dressed the loom. It wasn't in the book; they let me guess, and I will do the same for you. Any ancient weaver only made this same mistake but once, and so did I.

148 strings, fifteen feet in length, lay on the floor in one big pile. On my left, a loom that hadn't seen a string in forty years.

Now what to do?

Back to Ms. Black, her book said there are just six ways to dress a loom. What she meant was just six ways to do step ten. I studied each intently, as if I knew what each word meant, and finally chose to do the one called "Threading from the front." Not to ignore the "Swedish method," mind you, or the "Beriau" which I could not pronounce.

Step 1. "Tie the lease sticks…" I had no lease sticks. Find out what they are and then make some.

Step 2. "Cut the warp chain loop where it lies between the breast beam and the harness." My warp was lying on the floor in a big heap. The next step was unclear. I would have quit except I knew that Indian maidens, who could not even read, could do this easily while in their sleep.

Surely a sixty-year-old man, who claimed to be an engineer, could not be daunted by so primitive a task. Believe me, it isn't easy, and yes, he surely can.

About a week had passed in which every waking moment was spent trying to decipher what came next.

"*Insert the warp hook…*" Had to make one first.

"*Draw the warp ends through the reed.*"

"Thread the warp ends through the heddles."
"Tie the warp ends to the warp beam rod."
"Loose the lease sticks from the breast beam."
"Tie the ends together now."
"Let the lease sticks lie in front and by the reed."
"Wind the warp onto the warp beam,
Keep the tension of each one just right."
"When you reach the end, tie off onto the cloth beam rod."

"It's not necessary to rough sley or transfer lease sticks with this method, though it may seem awkward when first done, practice will make sleying, beaming, threading go quite quickly till the job is done."

That's what it said. If you don't understand, I didn't either that first time. There were fifteen pages telling how to do each step just right. At last it seemed a miracle, but there before me sat a loom all fully dressed. When I pushed on a pedal, a shed opened like it should. When I pushed another, it would open up again.

I was so proud I like to burst. I felt like the time I pushed my first stick in a Tinkertoy, and then called for Mom to show her what I'd done.

I made some shuttles out of hardwood scrap, sanded, smoothed and polished them. For practice I used some triple strand wool from a ragbag. I learned to tweed and twill and herringbone. I made patterns, played with colors, then decided that a tapestry was what I really wished to do.

But tapestry is hard to do, especially on a little loom, because there is no room to work.

Spring was fast approaching when I cut my last work loose. I knotted off the warp in March and have not touched it since. I looked at it the other day; all that it needs is just to add some warp and have inspiration, and some time to make it work.

I bought some books on Navajos working with the wool, and studied them with care. I learned the story of their weaving, how the skills and patterns grew, as they struggled out of poverty in their homes in the southwest.

Their work was primitive, yet fine and beautiful beyond belief.

That summer I designed a frame of three feet wide and six feet long that is balanced on a pedestal, so I can tie it at an angle that I choose. My miner's legs would never let me squat like Navajos when working on a loom, outside, beneath the desert skies.

On my Mac computer, I experimented with design. Most Navajo can have a complex pattern in their head. I needed something more.

I found that the bilaterally symmetrical designs they often choose, did not inspire me. I liked to have enlarging figures, spaced in hyperbolic arcs, across a background made to balance with the perceived movement of their form. What that really means is if you can't make a straight line, then try to make the crooked one be as pretty as you can.

I went to "Just Dyelightful," a weavers' store in Colorado Springs; I'd heard that Judy there was as expert as they come. She listened patiently as I explained that I knew nothing of materials. I didn't even know what type of wool to use for warp, or what the weft should be. I explained that I had taught myself to spin and weave with llama wool, had rebuilt wheels and looms, and made the other tools I needed. I told her of the loom I made, to use the Navajo technique.

She sold me wool for warp, and watched me choose the colors of my wool. She said, "I have a Navajo who comes here twice a week. I can set her up to help you with your patterns and your skills."

"I'd rather not," I said, "I've taught myself and come so far. I don't wish to be influenced by any other artist. I would rather struggle on and have a product that's unique."

She replied, "Well I'll be darned. For thirty years I've worked with yarn, I've never had a lesson in my life. I tried it once, but left before the morning session was complete."

Then she went on, "And now an old and gray-haired man tells me that he feels just like I did that many years ago. And has the same solution to his plight. I wish you well; please do not change. Sometime stop and bring your work, it will be something I must see."

I did stop in the following year; she was as proud of me as if I were a son. She showed my work to everyone. I could not leave it there, for I had promised it away. I still get flyers from her store. I am ashamed; I never have been back.

Design on the computer is an easy chore for me. I move the elements around, enlarge them and distort them, until a pattern falls in place.

I had no color printer, so I had to work directly with the wool. I would lay out skeins and swatches and try to picture in my mind how it would look. I would enlarge my printout, and count the warp each time and then compare it to my drawing, to be sure that it was right.

I envied native weavers, who did all of this and kept it in their heads, including warp count for each shade. There could be fifteen changes in each weft.

I had to make the beams on which to tie the warp, and then find the proper string to tie them solidly in place. There were shed rods to preserve the cross, and battens made to open up the shed. There was a heddle rod, and string to make the heddles that was strong and slick. I bought dowels and used broomsticks, nylon cord or cotton when the nylon was too slippery to tie.

My books about the Navajo were useful, for they told me how the Navajo made tools from wood and fiber from the land. I bought some weaving needles, made a beating fork from walnut I had saved. I found my fingers were not flexible and skillful, could not pass a "butterfly" of yarn on through the shed like native weavers do. My shuttles are of oak, and none of Navajo design. They look more like knives with yarn to wrap the handle tight.

It takes a good eight hours just to dress the loom, and twine and bind the ends onto the beams so I can tension up the warp. At last I'm set to weave, I'm using all commercial yarn, single ply and all the same in size; the colors are consistent, and there is no pattern at the start.

This should be easy. Guess again.

My warp pulls out of shape; the edge begins to angle in. I take it all apart, retension up the warp and try again. The warp now shows on through the weft, a slack loop forms. I take it all apart and try again. So very slowly does my work creep up the loom. How many times reworked, with low spots filled? I hooked the weft I should have turned; the right side falls behind and won't catch up. When I am half way done I realize it is a mess. I continued on, I needed practice anyway.

So slowly does my skill develop, I am consumed. I can spend every waking moment at the loom, gulp down my meals, awake at five a. m., and on into the night, with not a civil word to anyone. Then springtime came and there was outside work to do. I had to break addiction to that loom, and stay away for fear I'd start again.

When autumn came and farming work was done, I dressed the loom to make another rug. This time I didn't fight the wool, but played with it instead. I found a radio station that played classical jazz. My shuttles whispered to the beat, I hummed along and slowly, oh so slowly a skill was born and grew.

Perfection of the skill will never come, I know. But with commercial wool, or sheep's wool from New Zealand like Melissa brought, the work is passable, but not unique. It is a thrill to realize an old man had learned how to weave.

Now to do a work of art, of llama wool, hand washed and carded, spun and woven and designed by me. I started after New Years Day in 1993.

On the computer, I had drawn Thistle's face, the llama with the chocolate wool. I blocked the colors so they could be woven on the warp with eight strings to the inch. I had four colors, grayish white from Frolic, tan from Cider and Thistle's brown, along with half a pound of black, a gift to me. I cleaned and spun the wool, made just enough to start.

Most carefully I dressed the loom, and started then

to weave. I have advanced to just above the eyes, and there it sits.

It's beautiful, but I don't know if I'll ever finish it. (I have not as of May, 2005.)

In '93 my world turned upside down.

Three funerals in two months, a loving wife for over forty years was dead. Grandpa and Grandma, always close to us, were also gone. I was an emotional wreck, could not relax, my body ached. Thank goodness for my children, they so patiently put up with me, and helped me through my strife.

By '95, I'd found another loving wife, who finished up the job of bringing me back to my former self.

Almost.

I've lost the confidence that I once had. I don't think I would learn to weave today; I'd find some vague excuse, and put it off, and off again. Now tears come to my eyes, without reason; they just do.

My looms sit in the basement, the warp on one all knotted at the reed. From the other, Thistle's eyes stare out at me, his head stops at the eyebrows, he has no ears and without ears, a llama has no life. He has been buried under boxes, furniture and such, that accumulate when two homes become one.

Will I ever weave again? I honestly don't know.

The yarns of many colors are still piled high upon the shelf, and bags of llama wool are stacked beneath. The shuttles and the needles are still scattered on the table that holds the sketch of Thistle, all checked and marked with squares. My weaving books were on the shelf; until I brought them down to find the words that I'd forgotten in the last four years.

Maybe later, when I cannot see this damn computer screen, I'll take a shuttle in my hand, regain the touch, the feel of llama wool made into yarn, slipping smoothly through my hand.

My weft will tension up just right, my warp will all stay straight. I'll have my hands caress the cloth, and do again the things I loved to do when I was newly old.

Maybe someday when I'm gone, a grandson yet unborn, will fall in love with Grandpa's things. He'll save them from the trash man, and fix them up and learn to weave like Grandpa did, so many years before.

jimmyj 2/22/97

Chapter 11. Deeper reflections

In the last two chapters, I spoke of activities, the fun and games. Now I would like to delve a little deeper and examine some of my thoughts about life.

The Games, the Games

The Romans had it right, a democracy they claimed. All men were equal and a senate ruled, and made the laws, appointed one to be their leader. They set out to civilize the world they knew.

They gathered knowledge from the East, from Arabs and Greeks, Egyptians too, and brought it back to Rome, and soon, a power such as none had ever seen emerged. The armies traveled far, to Britain on the west, the Baltic on the north, east into Iran, to Egypt, Spain, Morocco; all the lands about the greatest inland sea.

Commerce grew. Trade and finance flourished, ships and caravans brought all the world's riches back to Rome.

But always on the fringe, there were barbarians, ready to advance and plunder, take the land, to rape and pillage and make dark the light that had come forth.

Not to worry, back in Rome, the games were on.

The lions and the bears, one hundred thousand watched the Games.

The Games, the Games. A hundred thousand watched the Games.

Two thousand years have passed, those same barbarians now have learned the way to get it right -- democracy -- where all are free and men can choose their leader if they wish.

Other nations lost their way, their grand experiments failed; the iron hands of dictators have grown soft. The world is complete, every nation is in touch. Information passes quickly from the far spots on the globe. Foot speed of the Roman's long surpassed by waves of light. And each of us can, if we wish, know more each day, than all the Roman legions ever knew.

But still barbarians lurk upon the fringe. Barbarians of the mind who wish to darken all the light we know.

But not to worry, for we have the games,

The Lions and the Bears, one hundred million watch and hear the football game.

The Games, the Games. A hundred million watched the Game.

The Roman aristocracy grew rich from commerce and built villas by the sea, with halls and baths, and gardens tended by their slaves. Food was brought from many lands, and feasting lasted through the night. There were orgies such as one had never seen. And still the caravans brought more; silks and furs, herbs and spices from afar to please the palate of the few.

In Britain, rumor said, the Scots were coming from the north. But do not worry; Hadrian will build a wall and stop the Scot barbarians there.

Let's get back to the games. Fourteen teams of four abreast, around the pylons until only one is left. Eight men killed, trampled by the feet of screaming horses, legs all mangled by the swords on chariot wheels. Tomorrow better yet.

The Christians and the lions, hear them roar? How they've been starved.

Tomorrow we will come back to the Games.

The Games, the Games. More than a hundred thousand watched the Games.

Men rich from hi-tech business, from investing in the market, build their villas by the slopes and shores, where they can play with skis and boats, and rock the night away in rooms with smoke of many kinds. They have their wilderness where they can go while they are young and strong, and then return to food from other lands all frozen fresh, with wine and margaritas, noise defined as music blaring forth. Electronic slaves control the light and heat and sound.

Violence and sex, idiotic talk shows, movie after movie on TV. Children watch for hours, instead of doing homework as assigned.

No one sees or cares to see barbarians of the mind. For soon we'll have a V-chip and a rating system for the tube. Then parents need not worry and can get back to the games.

Football, hockey, soccer games, and tennis, golf, and bowling games, basketball and baseball games, with hour on hour of pictures on TV.

And worse yet, the casinos take their tax.

One hundred thirty million watch the Packers and the Pats. All commerce stops, emotions take a wild ride, and everyone will watch the football Game.

The Games, the Games. A hundred thirty million watch the Game.

The Barbarians did not stop. There were Goths and Vikings, Huns and Vandals, Visagoths and Moors. Men were taken by the sword as if it was a game, and women then were taken as a wife.

The senate was corrupt, then gone. The Emperor was chosen by betrayal and deceit. The gates to villas then were barred, the Vatican was sacked, and the hired armies and the slaves turned upon their masters. Nero fiddled for the Games, and all was lost.

It took 1500 years to civilize barbarians that had come down from the north.

Can we hold off our barbarians, barbarians of the mind?

Look around you! Can't you see?

I can, but frankly I don't care. I really need to get back to the football Game.

A hundred thirty million and one more will watch the Game.

jimmyj 2/3/97

Alone

I sat bolt upright on the bed!

Wide awake in fright, I had reached out my hand and there was no one there.

Oh, my God, not again!

The tears burst from my eyes, ran freely down my chin. Oh, no, no, not again.

Then I awoke. The clock said four-fifty a.m. I tried to stop the tears, brushed them away upon my hand.

It's OK, I told myself. This is the way it ought to be. I'm alone by choice, not because the expected one has gone, and never will return. It's OK, OK, OK, OK!

It really is OK. Now stop your crying, Jim.

I rose and walked out to the kitchen in the dark, looked out the windows at the darkened sky.

Venus had risen; just the way it should, and glowed in promise of a day not dawned. The darkness was pervasive, the mountainous horizon still obscured. The lights of town are insignificant, man's vain attempt to chase away the night. I stood and stared from the cold room. What in the world had brought on this emotional storm?

I stood and cried, I couldn't stop.

Maybe if I make a pot of coffee, take a shower and shave, put on some clothes.

That doesn't seem to help. Maybe if I write it down, and try to think objectively, then I can stop these sudden bursts of tears. I dry my eyes, and wipe my nose, and sit here at the keyboard. Six twenty-six a. m., I wipe my glasses clean and write.

Here come the tears again.

Just yesterday we had gone to a funeral, my new wife Eleanor and I. Don Mullins, one whom I had never known well. Bobbie's Don, forty-four years a married man. To Eleanor, her own extended family, more like a brother than a friend.

As we stood in church, I held her hand, or she held mine -- it didn't matter which -- as tightly as we could, to give each other help and reassurance that each was not alone. For we had each been there, right where Bobbie stood, and we knew one of us would someday be right there again.

A voice sang out, so beautiful and clear, *"For he will raise you up on eagles wings...."*

Then it was done.

I left Eleanor in Denver, that she might help her friend recover from the shock. The time has been so short, she doesn't even know it's there. She will. And I came home with her son, Mike. Then I wakened in the night the way I did.

Go make yourself some breakfast, Jim. Stop your crying, you big kid. She'll be home soon to hold your hand. Or you hold hers -- it doesn't matter which.

Then, at least for now, neither needs to be alone.

jimmyj 2/28/98

So Many and Too Much

There was a time when I was young, a recurring thought went through my mind. If only I had manly strength, then surely happy I would be, I'd never ask for more.

But when the morning came, my puny, wimpy self remained. So then I wished for wisdom, for if I couldn't be strong, then surely wisdom would suffice. And when the evening came, I knew no more. What could I do?

I asked my Pa. He said "I cannot grant your wish for strength, but while you're wishing, shovel out manure from the chicken house and hoe the garden for your mom, then split some wood."

I asked my mother if she might, from her great store of wisdom, give me just a bit. She said "My son, the Lord does not allow a mother to bestow a gift of wisdom on her child. I can only give you love, it is my only choice. But here son, take this book and read. Don't waste your time in wishing for what will never be."

I chopped some wood and read a book.

As I grew older, then I wished to have the skill a ball to pitch. That's all I'd need. I asked a coach to give me some. He said "I've squandered mine away. I've none to give. But while your wishing, take this ball and throw it hard right back to me."

I wished that I had love, a wife perhaps, then we could raise a family and happy ever after we would be. I asked a friend, she said to me, "I've none to give, for you must first give it to me, then I may give it back, we'll see."

I chopped some wood and read a book, I threw it hard, and I gave my love.

And older still I thought that I could live forever, happily, if I were rich. Who can I ask? No one I see. And if there were, he'd probably say, "I've none to give, but while you're wishing, work and save and raise your family."

If only I would have a God to give me comfort, show the way, that's surely all I'd need. I asked the Father, he replied, "I've given my last one away to some poor beggar at my door. He needed it, you don't just now. But while you're wishing, why not pray?"

I chopped some wood and read a book, I threw it hard, I gave my love, and I worked and saved, and prayed.

I met a man who said to me, "My, how lucky you must be to be so strong and skilled, to have such wisdom and such love. To be so rich and have a God who gives it all to you for free."

How right he was.

Now older still, I cannot chop. I've lost my strength. It's hard to read, my memory's not always right. My wisdom slips.

I cannot throw, my skills are gone. I give and give my love away, but it keeps coming back to me. I have so many other things, so much that I don't need for happiness. So take some, please, I have enough.

And still I pray. He doesn't answer me.

The time will come, it's plain to see, when there'll be only God and me. I'll pray to him, he'll answer me.

He'll say, "Here is the key, but you must find your way to me."

The key is death.

     jimmyj 1/18/97

To be in Control, When that Day Comes

I hope to have the ultimate in self control, to look the doctor in the eye and then say, "Doc, let's stop this old baloney. No more chemo therapy, my friend, I am prepared to die."

Then to lay back on my pillow, to let the tears flow freely if I wish, and feel deeply in my failing heart and mind, a happiness so nearly lost by fighting to continue, when all the need to fight is gone.

Those of you who care may wring your hands, may even say, "He's lost his mind, so don't listen to him, Doc."

Please don't.

If such a thought should cross your mind, I'll make the statement once again, so listen up! Don't fret the details, friend. I am prepared to die.

Why not?

I've lived the lives of many in my time.

A boy raised on a farm. I learned the basics, about God and family, about work and animals, about life and death. A life that many never go beyond, it could be life enough.

I've been a student all my days, a life I love. There always is another thing to learn; there isn't time to study all a man should know. My books and magazines stack up, unread. It's hard to set each one aside as my mind loses the ability to quickly comprehend.

I had another life, a life of sports. I learned how to play the game, to know the rules and never cheat to win. When to lead and when to follow, take defeat with gritted teeth, congratulate and grin.

Now, when my meager skills are gone, I am privileged just to watch and cheer, to make believe I'm young again and have more skill and knowledge of the game than I ever had back then.

To be a husband and a father is a life all of its own. This is one that's not so easy. It took two of us to try and chart a path through eccentricities of parenting. It is with a lot of apprehension that I write these paragraphs.

Did we do it right? We didn't know, I don't know now. We weren't always successful sending children into the world. I have worried, sometimes patted myself on the back. Sometimes I feel so guilty, I could cry.

Now at my age, I'm thankful for our family. I cannot express the pleasure that my kids all bring to me.

All in all, we were quite happy. I'll let you judge if we did it right; we did the best we could. This life of mine continues; I am prepared to let it go, inconclusively, I know.

The life of a day laborer is one I thoroughly enjoyed -- to work each day, all day and every day. There is a feeling of contentment to be dog tired when the whistle blows, to quaff a beer, then eat a hearty meal and fall in bed, only to wake and start it all again.

To see construction that I'd done, a tunnel where one had not been, a landscape or a garden where only weeds had grown till then. This was a life, a good life and the only life that's known to many men.

But it was not for me. I've had the privilege to become an engineer: to plan and draw and calculate, to use the tools of math and science, then to see that which had been a dream, emerge in solid form before the world.

There is no pleasure or no satisfaction like that of seeing something new, to see some process used, to see a task made easier by engineering principles applied.

If you are once an engineer, you will always be an engineer. All other lives a man has had step back in deference to that one.

I've also had the privilege to be a manager of men. At first a few, and then some more, and them some more again. The group grows exponentially as you begin to manage men who manage men who manage men.

My ideas were examined, prospects were explored, construction was designed and mines were mined. Boys became men, students became professionals, apprentices to master craftsmen and much more -- did I really manage that?

It was hard to live and hard to leave, now it is gone. An experience few can claim.

Then I decided to back off, retire from the pressures placed upon myself by myself, and a job careening out of control. I became a farmer once again. With a little effort, I grew anything. I designed and installed landscapes, built a business up from scratch, turned it over to a son, so I could loaf again.

Did I have a religious life? Not really. When I was young I was conditioned to obey, so I went to church on Sunday because Mother said. I seldom miss a Sunday mass, I enjoy some mature thought and it is the proper thing to do. If my prayers for much produce a little bit, I can accept that as enough and thank my God.

I've studied hard, not only of my church, but others too. I've studied all the ancient ones, the gods of Greece and Norseland, Zoroaster, Buddha and Confucius. I learned how totems disciplined the stone aged tribes. I examined how the Hebrews claimed the faith; how Mohammed redefined the role of man and why Luther felt he must reform.

I know the struggles of our church, how time has tempered dogma, how the social structure changed. I was at one time very active, now I ride along and try my best to keep the faith. God has been kind in many ways; I will not ask for more. I am at peace with Him.

Now I live an easy life, doing almost as I please. I help out on the farm, a life I love. I choose to travel little; I choose to join no groups or clubs. I choose to spend my time at a computer, writing stories of the life I lived. I often take a noontime nap, to refresh my aging mind.

Don't fret for me when brass time comes, my shift is done. I've been everywhere I wished to go, done everything I wished and sometimes more. So let me step out of the way and let you continue on. There is no need that I forever keep control. We'll meet again, someday, somewhere.

jimmyj 8/1/99

On September 12, 2001 I was privileged to speak at the Optimist Club breakfast meeting. This happened to be the morning after the terrorist attack on Washington and the World Trade Center. The world was in shock as we all attempted to rationalize the events of the previous day.

After my usual lighthearted presentation, I observed that, like it or not, we had become engaged in a religious war. I noted that religious wars are always long lasting and inconclusive. I offered the following observation, which I had penned in 1998. It brought our discussion to a hushed and solemn close.

About God

Time is but an instrument of Man. A light year an inadequate concept to measure space.

Space is really undefined.

The atom, with its energy and mass, are only vaguely understood.

The smallest element of life must come from life, Man cannot start life anew.

It is obvious that space and energy and mass exist, and change with rules most absolute.

Rules that Man, with all his wisdom, cannot make. He can only theorize on what is true and what is not.

Is there a God?

Of course there is!

Someone made the rules, so energy and space and life may all exist.

How arrogant of Man to think he might define a God, a God with power so immense.

We must be humble before God.

jimmyj 2/14/98

Chapter 12, Media and the News

Let me change the subject. The modern information age provides a gem now and then. Witness this experience:

You remember March third, 1999, don't you?

That was the day Monica Lewinsky revealed all to Barbara Wolters. They say the network valued the programmed interview at thirty four million dollars, and it kicked off the sale of a book that will be read by multi-millions of people.

So I was ready to watch, with my zapper in hand, to see what else was going on in the world during the commercials, and there were many. In fact there was one on Channel Seven, so I zapped down a notch to PBS on Channel Six.

I guess it was the music, a soft clear, feminine voice singing, "I'll be seeing you -- in all the old familiar places -- that this heart of mine embraces -- all day through."

A rather amateurish introduction to a program called *"The Forties,"* and more music, always music, "Till then, my darling please wait for me...," the perfect harmony that could only be the Mills Brothers.

The 1940s: I was twelve years old when it begin, and grown up when it ended.

Music, always the music. As a grandmotherly Kay Starr said, "We knew all the words, always. I still do." And so do I.

I watched the program intently. It was a very emotional experience, I forgot about Monica until the first half-hour segment was over. When I zapped back to Channel 7, another commercial was on, so I returned to the "Forties" for the next hour and a half.

The 40s, how poor we were. We fortunate ones had a radio, and heard of the bombing of Pearl Harbor. Only later would we see pictures in a seldom-purchased Sunday paper picked up after church. Later yet, the Movietone News at the theater, or in high school special assemblies. But it was the radio, and the letters home from the many in the service of their country, that kept us informed of the

War.

President Roosevelt had been trying to warn us of the impending danger from a man called Hitler, but even he was not aware of the horrible sneak attack planned by Japan, the Land of the Rising Sun. December 7, 1941, wave after wave of bombers smashed our fleet anchored in the beautiful harbor at Honolulu.

Roosevelt had pushed through a lottery for a draft of men to build the armed services in 1940. He organized our country as free men volunteered to fight, and women took their places at the workbench and on the tractors of our land. We were woefully weak when the war began, but great men stepped forward to lead us: MacArthur, Eisenhower, Patton, Nimitz, Jimmy Doolittle and many more. Everybody's brother was in the war, somehow, someway.

And always the music, I'll Get By, The White Ciffs of Dover, I'll Be Home For Christmas, and Sentimental Journey. The Dorsey brothers, Bob Crosby, Glenn Miller and Johnny Mercer led the big bands.

We listened to the gravely voice of Churchill, the emotional fireside chats of Roosevelt, and the "buck stops here" determination of Harry Truman.

By May of 1945, we had victory in Europe. We dropped the Atom Bomb on Japan in August and the war was over. There was dancing in the streets.

GIs came home and went to college, got married and began to raise the Baby Boomers. The Big Bands dismantled in the late 40s, and only Lawrence Welk was able to make a transition to television and even today is the only replay of the popular swing music of that time. The *Greatest Generation*, as Tom Brokaw called it, had become history.

The program was over when I remembered why I was watching TV this night. I zapped back to Channel 7 and listened to the sickening voice of Barbara asking silly questions to a giggly Monica.

Seventy-one million Americans, whom the Media kept telling us were not interested, had watched the sorry spectacle about a sorry girl and a sorry president. What a contrast to the great generation of the 40s. What a sad and sick comparison. What a sad commentary on our instant information age, the people who present it, the people who watch it and the people who make the news.

jimmyj 3/99

While I'm panning the Media, let me continue with this observation:

This morning on the national news again, "There are a crew of forty standing by to report on the expected birth of septuplets in a small town near Des Moines, Iowa."

And on and on, until I turned off the TV in disgust.

There were thousands of reporters and their techno crews, a million exclusive interviews. A normal birth would still be two and a half months away.

How I pity the parents. Their lives will be more affected by how they are treated by the press and the publicity than they will by the birth of their babies.

Please, God, don't let anything go wrong. It's a disaster already.

How different it was two thousand years ago when, according to Luke 2:7, "And she brought forth her firstborn son, and wrapped him in swaddling clothes, and laid him in a manger, because there was no room for them in the inn."

The Media?

Luke 2:10-11, "And the angel said to them, 'Do not be afraid, for behold, I bring you good news of great joy which shall be to all the people; for today in the town of David a Savior has been born to you, who is Christ the Lord.'"

TV, Internet?

Matthew 2:2 "...for we have seen his star in the East...."

The result?

Matthew 2:13 "...take the child and his mother, and flee into Egypt...." and 2:16 "Then Herod....slew all the boys in Bethlehem and all its neighborhood who were two years old and under...."

Now keep in mind that Matthew was writing this in retrospect, 75 to 90 years after the birth, and Luke was probably later. Mark didn't comment at all on the birth, and in John 1:14, "And the word was made flesh and dwelt among us."

Now let me write for you a modern day parody.

November 18, 1 BC

"This is Peter Jennings speaking to you from a small town in Galilee called Nazareth. Behind me you see the carpenter shop of a man named Joseph whose wife is expected to give birth to God in about thirty-five days.

"Let me try to push my way into the face of Joseph, who is attempting to complete a cabinet for Ann, a relative, but has been unable to clear the reporters out of his way. In desperation, Joseph has just taken a swing at Tom Brokaw with a corn broom. I will have an exclusive interview with Tom later.

"There seems to be some question about the parentage of the expected child. I have an interview with a neighbor, the wife of a camel driver, who has been peeking from behind her veil at Mary's house for many years. She reportedly has seen someone, who -- unbelievably was not walking on the ground -- enter the home, and heard a voice say, 'Hail Mary, full of grace.' We are editing a home video of the activity, which you will see exclusively on NBC on the Ten O'clock News.

"I have with me a man who has made sandals for Joseph. 'Sir, do you expect that the sandals will last all the way to Bethlehem? If not, why not?'"

"Most assuredly they will, the soles are of the finest buffalo butt, and the straps are pure braided palm fronds."

"That was an exclusive interview with Nike Reebok, a sandal maker here in Nazareth. We will follow this important development on the walk to Bethlehem.

"Let me interrupt for coverage by CNN from Bethlehem. Here is Wulf Blitzer, who has left his post covering Caesar Augustus for this assignment. Can you hear me, Wulf?"

"Yes, Peter. It is widely reported that Joseph does not

have a reservation at the inn. I have here the innkeepers' eight-year-old son, as soon as he puts down his armload of camel chips. 'Son, does your father have a reservation for Joseph and Mary for December 25th?'"

"I don't know. I can't read."

"There you have it. From the Inn at Bethlehem this is Wulf Blitzer for CNN."

Jennings: "I will be right back after this message"

"Does your camel have sore knees? Come to John Elway Camels. Newest and best from the Sudan, in both two- and four- knee models."

Jennings: "Now to Afghanistan, where a reportedly wise man is at his telescope looking for a star, which he says is ... one moment please, this breaking just now, live by special report! From the palace in Jerusalem, where Herod is holding a press conference."

Sam Donaldson: "Could you tell me why you have chosen to cut off at age two instead of two and a half?"

jimmyj 11/18/97

Amen!
jimmyj 4/02/2005.

Chapter 13, Epilogue, 2004

We were approaching Climax from the north, on our way to Buena Vista. I was unable to drive, so my second wife, Eleanor, had taken me to Vail to see my surgeon. My first wife, Mickey, the mother of my children, had died from breast cancer in 1993, and Eleanor's husband, an old Climax co-worker had died in late 1989. Eleanor and I were married on July 1, 1995.

There was a roadside sign, "Beef Jerky for sale"

"That sounds good," I said. "Why don't you pull in across from the gatehouse?" She agreed.

She drove off the highway to park at almost the exact spot where I had stood in 1971 and debated with myself about the requested move beyond the Glory Hole to Western Operations in Golden.

The gateway to the mine was in the same place, but that was about the only thing that hadn't changed. There wasn't a trace of the town that once had existed. The hotels, schools, offices and many other buildings were gone. The mill buildings and open pit shops were painted an indiscriminate tan, replacing the active galvanized look. The hustle and bustle of mining and milling had become a deathly silence; there was no activity at all.

I searched for, but could not find, a raven floating on the updraft of the wind at the edge of the Glory Hole.

An irregular berm had been built along the east side of the highway, and it was landscaped with shrubs and evergreens. It appeared to be an attempt to hide the operation, as if it were something to be ashamed of.

The Glory Hole now extended from the top of Mount Bartlett to a series of open pit benches that had removed much of Ceresco ridge. I wondered if anyone knew exactly where the Continental Divide that once followed the ridge was now located?

Eleanor parked the car and came around to my side to give me the walker so I could stand and stretch my cramped legs. She had to help lift my legs out of the door; I swore that someday I would learn to walk again. Four major back surgeries had left a six-inch titanium plate bolted to my cervical spinal column and a neck brace supported my head. I was a little shaky, but I guess it was better than the alternative.

I couldn't help but wonder if I should have carried one less buzzy through the stopes, placed one less twelve by twelve timber cap overhead, thrown one less bag of dynamite or staggered through one less stench of powder smoke.

The cute little gal at the jerky stand put down her magazine and asked, "Can I help you?"

She looked closely at me, "Hey, you're the guy who wrote the book, aren't you? You stopped here last year and gave me one."

She continued, "Is that really the way it was?"

"Pretty darn close," I said. "I'll take one of the large bags of regular jerky." I handed her a ten-dollar bill. "Keep the change."

She smiled and said, "Thank You!"

I recalled that on my first day in the mine, I earned $10.44 for eight hours work.

Eleanor helped me back into the car and I said, "Well, this is the place I made my fortune."

She walked around to the other side and got in.

"What you really mean," she said, "Is that this is where you qualified for Social Security."

We drove in silence from "This Place" toward Buena Vista.

A lump formed in my throat as recalled *The Odyssey of Homer:*

Breathes there a man with soul so dead
Who never to himself hath said?
" This is my own, my native land!"
Whose heart hath ne'er within him burned
As home his footsteps he hath turned
From wandering on a foreign strand!

Home? My home? After seventy-six years?

The Wisconsin of my childhood has long been relegated to be the place I came from fifty some years ago.

Once it had been the Glory Hole at Climax, but that is becoming just a fond and distant memory.

Now there is no doubt. Home is the farm in Buena Vista.

jimmyj 3/22/2005

Appendix

A glossary of mining terms as used at the Climax Mine.

adit   The main tunnel, the horizontal entry to the mine.
bazooka   A short piece of two-inch pipe with a compressed air hose and a water hose attached. When turned on, it sprayed a mist of water and air to settle dust.
blasting stick   A 3/4"x1 1/2" straight grain, wooden stick up to 20 feet long, which was used to make bombs for blasting hangups.
bomb   A bag of powder tied on to a blasting stick.
brass   1. A round brass fob with the employees work number stamped on it. 2. management.
buzzer   Used as a signal to indicate the start of a shift of work.
buzzy   Air powered rock drill with an attached, in line air cylinder feed leg. A stoper.
car cleaner   1. The machinery at the ore bins which had a rake on extendible hydraulic arms. It scraped the sticky ore from car bottoms. 2. The car cleaner operator.
controller   A hand operated device to control the amount of electrical power applied to locomotive motors, thereby determining the speed.
dash   Slang for slusher drift, from the written name. For instance "120 S - 7", spoken as "one twenty south <u>dash</u> seven".
diggers   Work clothes.
drawhole   1. The 4' by 6' transfer point from a slusher drift to a muck car. 2. An opening where you could dump ore into an ore pass.
drawpoint   A specific location where ore was available for production.
drift   A horizontal mine opening, tunnel.

fingers The forty-five degree openings from a slusher drift to the cave, through which the ore descended.
Fire in the hole! Always the warning cry indicating an explosive shot was about to be set off.
footwall 1. Generally a reference to the sloping underside of an ore body. 2. A haulage drift that traverses the footwall of the ore body.
Fremont Trading Company, The concessionaire for the store, bar and gas station at the summit of Fremont Pass.
Glory Hole The entire caved area above the mine.
glory hole An opening to the surface through which ore can be dropped into an underground mine.
guard Stand guard to keep people from walking into an explosive shot. A very serious responsibility in a mine.
hand A worker.
hanging wall The topside of a sloping ore body, or slang for the drift in this position.
hangup A large rock blockage in the finger from the cave, an ore pass blockage
high rock A loaded mine car with a rock so high it might strike overhead timber.
highball Fast speed, or a signal for fast speed.
lamp belt A wide belt that carried the battery for the cap lamp and the mine safety canister.
make it hot Set off the dynamite, by spitting the wick, or throwing the switch.
mantrip A train of special cars for hauling men into the mine.
MoS2 Molybdenum disulfide, the mineral, commonly called Moly.
muck boss Shifter on a production crew.
muck crew Ore production crew.
muck stick Shovel.
nose cable A nose cable pulled the scraper forward with a load of muck, which fell through the drawhole and into the haulage car below.

on the brass. Wide open, as fast as a mine locomotive would go.
One Switch The control center for underground train traffic.
Outside Anywhere except in the mine.
paste Loose dynamite usually held in the hand or dumped out of the bag.
peanuts Small size, free running ore.
pie can Lunch box.
portal The surface entrance to a mine via an adit or tunnel.
powder Dynamite, blasting powder or Ammonium Nitrate treated with fuel oil. Any explosive.
Primacord Ensign-Bickford Primacord, a detonating fuse had instantaneous ignition the length of the fuse and it would detonate most explosives.
primer A detonator, inserted in a stick of powder became a primer.
raise A mine opening extending upward.
ribs The sides of a mine opening, the walls.
shifter The immediate management boss, shift boss, reports to a foreman.
Slop Chute The beer bar and restaurant of the Fremont Trading Company at Climax.
slusher drift The drift from which the ore from the cave is scraped to the drawhole to fall into the haulage cars.
spit the wick Light the fuse.
spitter A cardboard device, about pencil size slipped on a wick to light it. A protruding wire ring was jerked forcefully to cause a sparking which lit the fuse powder.
spotted A mine car placed in the proper position for loading.
stopes The preliminary openings mined to facilitate a cave.
tail cable Pulled the scraper back in the slusher drift.

tallied Counted.
tally out Counted outside.
the back Overhead, the ceiling, inside top of a mine opening.
the windows Where the miners picked up their brass and daily work instructions.
This Place A miner's reference to the whole Climax Mine.
Whiskey Tech Slang, Wisconsin Institute of Technology.
whistlepunk A motorman's helper, early day trains were signaled with whistles.
wick Black wick, a dynamite cap igniting fuse, burned at a slow specific rate for safety and for timing hole detonation.
Wisconsin Tech Wisconsin Institute of Technology, now The University of Wisconsin at Platteville.

Sources.

The stories *To Illinois in '44* and *The Four Mile Ranch* were first published by Colorado Central magazine.

The photograph of Jim Ludwig is by Lauren McEvoy.

All other photos are from the author's collection.

A Brief Biographical Sketch of the Author

James J. (Jim) Ludwig, pen: jimmyj

Jim Ludwig, a Wisconsin native, came to Climax in 1950 as an engineering student from the Wisconsin Institute of Technology, now "The University of Wisconsin, Platteville." After a stint as a mine laborer, he became a contract miner. He returned to Wisconsin and completed his degree of Bachelor of Science in Mining Engineering.

Returning to Climax, he worked through the ranks of underground mine management to become Mine Superintendent in 1963. He began his general management career as the Assistant General Superintendent in 1965, and became General Manager in 1969. In late 1971, he moved to the Amax Western

Operations office in Golden, Colorado as the Manager of Mines.

Jim cut his managerial teeth while living in Climax. He was the last of mine management that would awaken in the night if the trains ceased rattling down the track, and to whom the sparking of the trolley poles was as reassuring as the northern lights are to Eskimos.

He requested early retirement in 1982 while he was Senior Vice President of Climax, in charge of operations, engineering and exploration.

He is the founder of the Pleasant Avenue Nursery, Inc. in Buena Vista, Colorado, a company that has developed the art of raising high altitude native plants for reclamation and landscaping. Now retired, he has seen the company become an important plant supplier throughout Colorado's high Country. He raises garlic, daylilies, peonies, and lilies for the nursery.

He is a remarried widower with an extended family throughout the United States.

He began writing as a hobby at a late age, generally to inform his children "who he really was," and then found an interest among his old mining friends for his accounts of growing up, while working in the mine.

In 1999, he published "The Climax Mine, An old Man Remembers the Way it Was" as a rambling narrative of the "good old days".

Ill health interrupted his writing career for several years. Recently he has resurrected many unpublished stories and essays, culminating in the publishing of this book.

"Jim Ludwig's *Beyond the Glory Hole* is a memoir of the life and times of a Depression-era farm-boy who became a mining engineer, general manager of the Climax Mine, and founder of a high-altitude tree nursery. Ludwig's tales are at times sad, often hilariously funny, and always insightful."

Steve Voynick, author
Climax: The History of Colorado's Climax Molybdenum Mine

"Jim's descriptive, lively, and highly personal writing originates from his varied life experiences growing up in Wisconsin and follows his career as a mining engineer in Colorado. Each chapter provides thoughtful, entertaining and often humorous accounts of his journey through life, and of the friendships he cultivates on his way to the "Glory Hole." The book is an enjoyable, thought provoking read."

Bob Renaux

Books published by Pleasant Avenue Nursery, Inc.
Mail order to:
Pleasant Avenue Nursery, Inc.
PO box 1669
Buena Vista, CO 81211

Phone order to: 719-395-6955, Fax order to: 719-395-5718

E-mail order to: pan@amigo.net or jamesjlud@amigo.net

Shipping and Handling: $3.50/shipment plus $0.50/item
 (1 book in bubble envelope = $4.00)

Beyond the Glory Hole, ©2005   Jim Ludwig $17.95
 A memoir of a Climax Miner
The Climax Mine, ©2000   Jim Ludwig $17.95
 An old Man Remembers the Way it Was

Payment must accompany order.
We accept Personal Check or Money Order
We accept Visa, Master Card, or Discover with order or over phone
Must have: Credit card #, expiration date MM & YY, Three digit security number from back of card, persons name on the card, and zip code of billing address for that card.

Payment must clear before shipment.

Shipment via USPS, book rate

 You may use order form on next page

Pleasant Avenue Nursery, Inc.
Publishing division
P.O. Box 1669, 506 S Pleasant Avenue
Buena Vista, CO 81211
Phone: 719-395-6955, Fax: 719-395-5718
E-mail: pan@amigo.net or jamesjlud@amigo.net

Purchaser _____
Address _____
City _____ State ___ Zip _____
Residence phone: ____ ____ _____
Business phone: ____ ____ _____

Ship to: ___ same as purchaser, or:
Name _____
Address _____
City _____ State ___ Zip _____

Item	Quant	Each	Total

<u>Beyond the Glory Hole</u>, ©2005 ___ $17.95 _____
A memoir of a Climax Miner
<u>The Climax Mine</u>, ©2000 ___ $17.95 _____
An old Man remembers the Way it Was
Shipping, $3.50 each Shipment _____
    plus $0.50 each Item ___ _____
Tax if applicable (CO 2.9%, Chaffee Co 2%) _____

Total (include with order) Ck #_____ _____
    or
Credit card: Visa, MC, Discover (circle one)
Number: _____
Expires: MM __ __ YY __ __
Security No. __ __ __ (back of card)
Card issued to: (name) _____
Card billing zip code _____
Signature _____